$500

The Velocity of Light and Radio Waves

The Velocity of Light and Radio Waves

K. D. FROOME and L. ESSEN

National Physical Laboratory
Teddington, Middlesex, England

1969

ACADEMIC PRESS
LONDON AND NEW YORK

ACADEMIC PRESS INC. (LONDON) LTD
Berkeley Square House
Berkeley Square,
London, W1X 6BA

QC
407
.F75

U. S. Edition published by
ACADEMIC PRESS INC.
111 Fifth Avenue,
New York, New York 10003

Copyright (c) 1969 By ACADEMIC PRESS INC. (LONDON) LTD

All Rights Reserved

No part of this book may be reproduced in any form by photostat, microfilm, or any other means, without written permission from the publishers

Library of Congress Catalog Card Number: 69 — 16495
SBN 12—242850—1

PRINTED IN HUNGARY

by the Akadémiai Nyomda, Budapest

Preface

Probably no constant is of more fundamental importance in physical theory and practice than is the velocity of light; this is reflected in the great effort which has been expended in its measurement. It seems to us, therefore, that there is a need to present this work in a single volume so that the various methods can be critically examined and compared. Our emphasis has been deliberately on the practical details of the experiments and on their gradual improvement leading finally to equipment which can be used for the measurement of distance. The major portion of the text will be readily understood by anyone with a general interest in the subject and the post-war work is described in sufficient detail to satisfy those having a more specialist interest and requiring a comprehensive treatment. Determinations made prior to 1900 are briefly discussed because of their historical interest and influence on subsequent experiments. At this point it was thought desirable to devote chapters to the nature of light and to a discussion of the basic standards of length and time in order to show that the uncertainty in the values of these standards should not so far have been a limitation to the accuracy of the measurements of velocity but that this position may not continue much longer. The errors of measurement are also discussed because although standard texts give a mathematical treatment this is seldom applicable to the velocity of light measurements in which the accuracy is found to be limited by systematic errors. The correction for the refractive index of the atmosphere is considered fully because of its importance in some of the post-war work and in geodetic distance measurements.

Velocity of light determinations made between 1900 and 1940 are described in two chapters one devoted to basically optical methods and the other to electrical methods not involving a beam of light. Attention is directed towards the work thought to be of greatest significance from the point of view of the experimental technique on its influence on the value generally accepted at that time.

The determinations made after the war from 1945 onwards took on quite a different character. Largely through the application of high frequency radio techniques the precision was increased enormously. This was clearly shown in the first post-war result which revealed a considerable error in the accepted value. It became possible to design equipment for the measurement of distance from the time of flight of a pulse of light or radio waves, the value of velocity being assumed. These methods and others designed specifically for the measurement of velocity are described in detail, emphasis being placed on those having the highest precision of measurement. A final chapter describes some new methods of measuring the velocity of light which are under consideration including one designed for the rapid measurement of distance exceeding 30 m.

As experimentalists keenly interested in the applications of science it is a source of satisfaction to us that the work described in this volume has led to a revolution in the measurement of distance. Relatively simple equipment has been designed to give results with an accuracy approaching 1 part in 10^6; and more elaborate equipment is becoming so precise that proposals are already being made to accept the value of the velocity of light as a unit of measurement, the unit of length then being derived from this and the unit of time — which is the most accurate of all physical constants.

March 1969

The authors are grateful to the National Physical Laboratory for permission to publish the following figures: 6.4, 6.5, 6.6, 6.7, 6.8, 8.1, 8.2, 8.3, 8.4, 8.5, 12.1, 12.2

Jo Aga Electronics Ltd., for supplying figures: 9.4a, 9.4b, 9.5.

Jo Tellurometer Ltd., for supplying figures: 10.2, 10.3a, 10.3b.

July 1969

Contents

Preface v

Chapter 1. Early Measurements
    1. Roemer 1
    2. Bradley 2
    3. Fizeau 3
    4. Foucault 4
    5. Other Optical Measurements Prior to 1900 5
    6. Electrical Methods of Measurements 6
    7. Summary of Results 9

Chapter 2. The Nature of Light
    1. The Corpuscular And Wave Theories 12
    2. The Wave Theory In Optics 13
    3. The Electromagnetic Theory of Light 15
    4. The Experiments of Hertz 17
    5. The Diffraction of Light 18
    6. The Nature of Light 18

Chapter 3. Standards and Accuracy of Measurement
    1. Definitive Standard of Length 20
    2. Definitive Standard of Time 22
    3. Application of the Definitive Standards 23
    4. The Setting Accuracy 23
    5. Systematic Errors 24
    6. The Refractive Index of Air 24
    7. Presentation of the Results 28

Chapter 4. Optical Methods 1908—1940
    1. Michelson's Determination In 1924—1926 31
    2. Michelson, Pease and Pearson 1935 32
    3. Karolus and Mittelstaedt 1928 33
    4. Anderson 1937—1941 36
    5. Hüttel 1940 39

Chapter 5. Electrical Methods 1900—1940
    1. Rosa and Dorsey (1907) 41
    2. Mercier 1923 45
    3. Summary of Values Obtained Between 1905 and 1940 . . . 49

Chapter 6. Cavity Resonator Measurements
    1. Theory of the Cavity Resonator 50
    2. Frequencies of Cavity Resonators 51
    3. The Quality Factors of Cavity Resonators 55

　　　　　　　4. Effect of Wall Losses on Frequency 55
　　　　　　　5. Essen and Gordon-Smith 1947 57
　　　　　　　6. Essen 1950 64
　　　　　　　7. Hansen and Bol 1950 72
　　　　　　　8. Cavity Resonator Measurements Proposed
　　　　　　　　　by Zacharias and Harrison 74

Chapter　7. Radar, Spectroscopic and Quartz Modulator Measurements
　　　　　　　1. Aslakson 1949 75
　　　　　　　2. Florida Survey 1951 76
　　　　　　　3. Other Radar Measurements 78
　　　　　　　4. Spectroscopic Methods 80
　　　　　　　5. Quartz Modulator Method 83

Chapter　8. Microwave and Radio-wave Interferometer Methods
　　　　　　　1. Introduction 86
　　　　　　　2. Froome's Determination of 1952 87
　　　　　　　3. Froome's Determination of 1958 95
　　　　　　　4. Florman's Radio-Frequency Interferometer 111
　　　　　　　5. The Microwave Interferometer of Simkin, *et al.* 113

Chapter　9. Modulated Light Beam Methods of High Precision
　　　　　　　1. Introduction 114
　　　　　　　2. Bergstrand's Geodimeter 114
　　　　　　　3. Geodimeter Results for c_0 120
　　　　　　　4. The Ultrasonic Light-Modulator 125

Chapter 10. The Tellurometer
　　　　　　　1. Introduction 128
　　　　　　　2. The Method of Measurement 129
　　　　　　　3. Velocity Measurements 132

Chapter 11. Summary of Results　　　　　　　　　　　　　　　　136

Chapter 12. New Methods under Consideration
　　　　　　　1. The N. P. L. Mekometer III 140
　　　　　　　2. Proposed Method Using Gamma Rays 145
　　　　　　　3. The Laser Beat Method 146
　　　　　　　4. The Light-Pulse Recycling Oscillator 148
　　　　　　　5. Sub-Millimetre Wave Interferometer 149

References 150

General Bibliography 152

Author Index 153

Subject Index 155

CHAPTER 1

Early Measurements

The idea that light travels with a finite velocity probably arose from an analogy with sound waves and mechanical vibrations. Galileo, born in Pisa in 1564, seems to have been the first scientist to consider the possibility of measuring its velocity. Observers stationed on opposite sides of a valley were each to be provided with lamps and shutters. One observer would open the shutter, the other observer would open the shutter of his lamp as soon as he saw the distant light, and the first observer would note the time between opening his shutter and seeing the other lamp. It is not known whether such an experiment was ever attempted, and it could not of course have succeeded, although the method was sound in principle.

Galileo contributed indirectly to the first measurement of the velocity of light by his application of the telescope to astronomy and the discovery of Jupiter's satellites.

1. ROEMER

Roemer, born in Aaruus in 1644, observed an irregularity in the time of revolution of a satellite of Jupiter (Fig. 1.1). The times at which it disappeared behind the planet were recorded throughout the year. The time of revolution of the satellite is approximately 42 h 28 min but Roemer found that as the earth moved in its orbit away from Jupiter the times at which the satellite

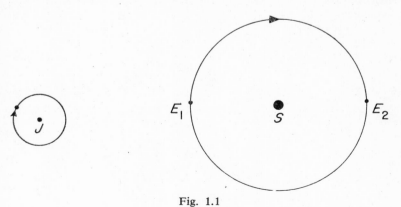

Fig. 1.1

entered the shadow lagged behind the expected times. He deduced that the extreme time lag corresponding to the minimum and maximum distances between the earth and Jupiter represented the time taken by light to travel across the diameter of the earth's orbit. He obtained a value for the velocity of about 214, 000 km/sec, the error being largely due to the uncertainty in the diameter of the earth's orbit. Later Delambre (1790) and Glasenapp (1874) obtained values of 986 sec and 1001·6 sec for the time lag and the mean of these values together with the present day value for the diameter of the earth's orbit ($2·99 \times 10^8$, km) gives a value for c of 303,000 km/sec with an uncertainty of about 2,000 km/sec.

Roemer's method would now be regarded most simply as an example of the change in the repetition frequency of an event with the velocity of the observer resulting from the Doppler effect. The period of the satellite changes by about ± 15 sec from the average values as the earth revolves round the sun, approaching and retreating from Jupiter at a speed which at its maximum equals the orbital speed.

2. BRADLEY

The next recorded determination was made by Bradley in 1726 (Bradley 1728) at Kew Observatory. He observed the star of Draconis and found that its position apparently changed during the year in a manner that could not be attributed to parallax.

Fig. 1.2

In order to study the problem further he had a new telescope erected at his home in Wanstead (U.K.) and there observed this apparent motion of a number of stars over a long period, finally arriving at the complete solution. To quote from his paper "Mr. B considered this matter in the following manner. He imagined CA (Fig. 1.2) to be a ray of light falling perpendicularly on the line BD; then if the eye be at rest at A the object must appear in the direction AC whether light be propagated in time or in an instant. But if the eye be moving from B towards A and light be propagated in time with a velocity that is to the velocity of the eye as CA to BA, then light moving from C to A while the eye moves from B to A that particle of it by which the object will be discerned when the eye in its motion come to A is at C when the eye is at B. Joining the points B, C, he supposed the line CB to be a tube inclined to the line BD in the angle DBC of such a diameter as to admit but one particle of light, then it was easy to conceive that the particle of light at C by which the object must be seen when the eye as it moves along arrives at A would pass through the tube BC ... and that it would not come to the eye ... if it had any other inclination to the line BD." He then proceeded on this basis to a consideration of the apparent movement of actual stars with the motion of the earth round the sun; and from the results of his observations deduced that the angle of aberration ϕ was 20·2" and that the ratio of the velocity of light to the velocity of the earth's motion in its orbit was therefore 10,210 to 1. This gives a value for c of 301,000 km/sec. Having established the theory Bradley then proceeded to confirm it by observations on other stars. It is easy to understand on reading this paper of Bradley's why Newton referred to him as "the best astronomer in Europe".

Bradley's method and that of Roemer are of particular interest because they give the velocity in a single direction. All the methods made on the surface of the earth give the average value along a go and return path. The fact that the values are the same within the limits of experimental accuracy is of great theoretical interest.

3. FIZEAU

The problem of measuring the time interval occupied by light in travelling a relatively short distance on the earth's surface was first overcome by Fizeau (1849) and in doing so he introduced a principle of fundamental importance in the field of measurement. Instead of trying to measure the short interval occupied by one return journey of the light, he arranged for a regular repetition of the journey and observed some parameter, in this case the intensity of the light returned, which reached an optimum value when the time of repetition agreed with the time of travel. The time measurement was thus replaced by the measurement of the rate of repetition or frequency, which is a far easier technical problem.

A source of light S (Fig. 1.3.) was focused by L_1 on to the rim of a wheel having 720 teeth. The light beam passed through a gap between the teeth, was made parallel by the lens L_2, focused by L_3 on to the convace mirror M_2 and reflected back through L_3 and L_2. It again passed through the gap and was transmitted through the half silvered mirror M_1 to the eye piece O. If the wheel rotates at such a speed that the light on its return strikes a tooth the intensity observed at O is a minimum, but if it strikes the next gap the

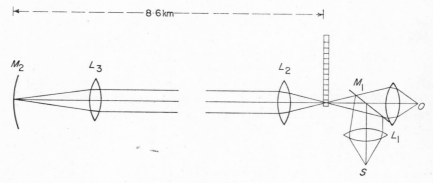

Fig. 1.3

intensity is a maximum. The first minimum was obtained at a speed of 12·6 turns/sec. The time occupied by the light to travel 17·266 km was thus $1/12·6 \times 1440$ sec giving a value of 313,000 km/sec for c. Doubling the speed of rotation would result in a maximum intensity and it is clear that the precision of setting increases with the speed and the number of teeth that are by-passed. Fizeau however never reported the details of his experiments apart from a single result which was stated to be the average of 28 measurements. This value was given as 70·948 leagues of 25 to the degree/sec, corresponding to the above value in modern units.

4. FOUCAULT

The second successful determination was made by Foucault in 1862 by means of a rotating mirror. Light from the source S (Fig.1.4) is reflected from the rotating mirror M_1 through a lens to the concave mirror M_2. It is returned along the same path to M_1 and, if the mirror is not rotating, it continues back to the inclined plate P from which it is reflected to O. If the mirror is rotating the image is displaced to O^1. The velocity is calculated from the distance OO^1, the speed of rotation of the mirror and the distance M_1M_2. In order to reduce diffraction effects the source was made in the form of a grid of parallel lines 0·1 mm apart. A disadvantage of the method is that as the distance M_1M_2 is increased the intensity of the image decreases rapidly, as $1/D^3$, since the reflecting mirror can be effective during only a small angle of its rotation. This was overcome to some extent by the use of a more elaborate optical system. Foucault used a chain of five concave mirrors. The mirror was driven by a compressed air turbine and its speed

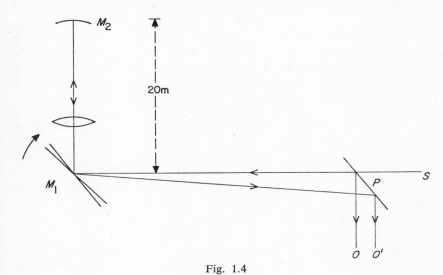

Fig. 1.4

was determined by a stroboscopic method. The speed used was 500 turns/sec and with an effective light path of 20m the deflection was 0·7 mm. The value obtained was 298,000 km/sec with an estimated uncertainty of 500 km/sec.

5. OTHER OPTICAL MEASUREMENTS PRIOR TO 1900

The methods of Fizeau and Foucault were used with minor improvements in many subsequent measurements. Cornu (1874) using a toothed wheel and a light path of 10·3 km made a series of over 600 measurements in 1872 and obtained a result in reasonable agreement with that of Foucault. He was encouraged to repeat the experiment by the Council of the Observatory of Paris in order to obtain a value accurate to 1 part in 1000 for use in connection with observations of the transit of Venus on December 9th, 1874.

He employed a number of toothed wheels of different designs and a path length of nearly 23 km. The precision of measurement was still low, the scatter between individual measurements being about 3%. He relied on taking the average of a great many observations weighting the results in a manner he considered appropriate. He thus obtained a final value of 300,400 km/sec for the velocity in a vacuum. The dangers inherent in this method of averaging widely scattered observations is illustrated by the fact that Dorsey (1944) made a very thorough study of these same observations and concluded that they gave a result of 299,900 km/sec.

The determination of the velocity of light was also considered of national importance in the USA and in 1879 Congress made an appropriation for the work and gave Newcomb, the astronomer, the responsibility for doing it. At this time Michelson was preparing to make an independent determination and it was arranged that he should assist in Newcomb's work. The main source of error in Foucault's method rested on the small displacement of the light image. This could be improved only by increasing the length of path but it was difficult to do this without losing too much light. Newcomb used two distances of 5·1 km and 7·4 km and many different speeds of rotation for the mirror. The mirror was a square steel prism with polished plated faces. The speed of rotation was set and controlled to give a fixed image in the receiving telescope first on one side of the point of zero deflection and then on the other. The results were less scattered than those of previous workers and after suitable weighting and averaging the value of 299,810 ±30 km/sec was obtained. Newcomb suggested a number of improvements which were later put into effect by Michelson.

Michelson (1880) carried out in 1878/9 the first of his long series of measurements. One innovation was the control of the speed of the mirror by means of a tuning fork. The image displacement of about 130 mm was measured with a micrometer and recorded to the nearest 0·01 mm. The average value of 100 determinations was 299,910 ± 50 km/sec. Later in 1882, partly because of the discrepancy between the above value and Newcomb's value, Michelson made another series of measurements using Foucault's method and obtained the value 299,853 ± 60 km/sec. His later work will be described in Chapter 4.

6. ELECTRICAL METHODS OF MEASUREMENT

In the latter half of the nineteenth century interest in the value of the velocity of light arose in a quite different field of science. The basic laws of electricity and of magnetism had been developed along independent lines leading to different definitions of the quantities concerned. It had been established experimentally by Coulomb in 1785 that the force between two charges q_1 and q_2 varied as the product of the charges and inversely as the square of the distance d between them. This law can be expressed in the form

$$F = K_1 q_1 q_2/d^2 . \qquad (1.1)$$

As the units of force and distance were already established independently it was legitimate to put $K_1 = 1$ and thus define the unit of charge as that charge which exerted a force of 1 dyne on a similar charge at a distance of 1 cm. The units of other electrical quantities such as capacitance, current and voltage could then be derived from that of charge. This system of units is known as the electrostatic system. Coulomb also established that the law giving the force between magnetic poles was similar to equation (1.1) Isolated magnetic poles cannot be obtained but a long bar magnet behaves as though it had a magnetic pole of opposite sign at each end. The repulsion between two similar poles could therefore be investigated and although the measurements were not of high precision they led to the inverse square law

$$F = K_2 \, m_1 m_2 / d^2 \tag{1.2}$$

where K_2 is a constant. The close connection between electricity and magnetism was shown by experiments, particularly those of Ampere and Faraday and it is possible to derive the units of other electrical quantities from that of unit magnetic pole defined by putting $K_2 = 1$ in equation (1.2) This system of units is known as the electromagnetic of c.g.s. system. For the solution of practical problems, such as the design of submarine cables, it was necessary to know the relationship between the units, and Maxwell for example regarded it at the time as the most important measurement in electrical science. The relationship can be determined by measuring the value of an electrical charge, a voltage, or a capacitance for example, by two different methods, giving the result in the different units. The relationship, which we can call c, is obtained as Q_s/Q_m, V_m/V_s or $\sqrt{C_s/C_m}$ where the suffix s denotes that the charge is expressed in electrostatic units and m in electromagnetic units. The first measurement was made by Kohlrausch and Weber in 1857 (see Maxwell 1904). They measured the charge on a capacitor, using a "Leyden Jar". The value for its capacitance was obtained in e.s.u. and this was compared with the calculable capacitance of a sphere and its voltage which was also determined in e.s.u. by means of an electrometer. The charge is the product of the capacitance and voltage. The charge in e.m.u. was then determined by discharging the capacitor through the coil of a galvanometer. The transient current gives to the magnet of the galvanometer a certain angular velocity. It swings through an angle θ at which it is brought to rest by the opposing action of the earth's magnetic field. The extreme deviation θ of the galvanometer gives the value of Q in e.m.u. from the formula

$$Q = \frac{H}{G} \cdot \frac{T}{\pi} 2 \sin \theta/2 \tag{1.3}$$

where H is the intensity of the horizontal component of the earth's field, G

is a constant of the instrument and T is the time of a single vibration of the galvanometer.

They thus obtained a value for c of 310,740 km/sec. Even before this value was obtained Faraday had tentatively suggested in 1846 that light consists of electromagnetic radiation and this idea, confirmed by experiment, became one of the foundation stones of Maxwell's electromagnetic theory of light. The merging of electromagnetism and optics is one of the great events in science. It is perhaps worth recalling Maxwell's comments on two of the experimenters who made it possible. "The experimental investigation by which Ampere established the laws of the mechanical action between electric currents is one of the most brilliant achievements in science. The whole, theory and experiment, is perfect in form and unassailable in accuracy; and it is summed up in a formula which must always remain the cardinal formula of electrodynamics.

"We are led to suspect what indeed he tells us himself, that he discovered the law by some process which he has not shown us and that when he had afterwards built up a perfect demonstration he removed all traces of the scaffolding by which he had raised it. Faraday, on the other hand, shows us his unsuccessful as well as his successful experiments, and his crude ideas as well as his developed ones, and the reader, however inferior to him in inductive power, feels sympathy even more than admiration and is tempted to believe that if he had the opportunity he too would be a discoverer. Every student should therefore read Ampere's research as a splendid example of the scientific style in the statement of a discovery, but he should also study Faraday for the cultivation of a scientific spirit."

In the following years a number of determinations of the ratio c were made but the accuracy achieved was not high. One measurement made in 1906, however, was as accurate as the direct measurements of the speed of light at that time, and this will be described in more detail in Chapter 5. Electromagnetic radiation predicted by Maxwell was produced experimentally by Hertz and he conducted experiments which showed that they travelled with a finite velocity. A value was first obtained by Blondlot. The spark discharges then used as sources radiate energy over a wide band of frequencies but a particular value can be selected by means of a tuned circuit consisting of an inductance and capacitance; the resonant frequency being given by the formula

$$f = 1/2\pi \sqrt{LC}. \tag{1.4}$$

The selected frequency was transmitted along a pair of parallel wires (Fig. 1.5) and reflected at the far end. If the amplitude of the incident wave is

$$E_1 = E_0 \sin(\omega t - 2\pi x/\lambda) \tag{1.5}$$

1. EARLY MEASUREMENTS

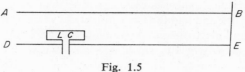

Fig. 1.5

and that of the reflected wave

$$E_2 = E_0 \sin(\omega t + 2\pi x/\lambda + \phi) \tag{1.6}$$

where ϕ is the change of phase on reflection and x is the distance measured from the point of reflection, the resultant amplitude is

$$E = E_1 + E_2 = 2E_0 \cos(2\pi x/\lambda + \phi/2) \sin(\omega t + \phi/2). \tag{1.7}$$

This represents a system of stationary waves the amplitude being zero at positions for which

$$2\pi x/\lambda + \phi/2 = (2n+1)\pi/2 \tag{1.8}$$

and $2E_0$ at positions for which

$$2\pi x/\lambda + \phi/2 = n\pi. \tag{1.9}$$

Successive nodes or antinodes are thus spaced at distances of $\lambda/2$ along the wire and λ can be determined. The velocity is then simply $f\lambda$. In this connection Blondlot refers to the work of Sarrasin and la Rive who had established that the wavelength along parallel wires was the same as that in air. The frequency was determined from equation (1.4), the inductance being calculated from the dimensions of the circuit and the capacitance measured by Maxwell's method. It is now known that many of the assumptions involved in this experiment are not strictly true and in the circumstances the result was reasonably accurate. Thirteen different frequencies between 10×10^6 and 30×10^6 Hz were employed, the total spread of the values of velocity was 5% and the average value was 297,600 km/sec. Blondlot concluded that the value agreed with the velocity of light and with the ratio of e.s.u. and e.m.u. units within the accuracy of the experiment. He added cautiously that it would be rash to conclude that the vibrations are of the same nature although the results are favourable to such a hypothesis.

7. SUMMARY OF RESULTS

These early measurements established a number of results of fundamental importance. Within their accuracy of about 1% the velocity in one direction along a path across the earth's orbit, and thus almost free from the earth's environment, was the same as that of a go and return path on the earth's surface. The velocity of light was the same as that of the radiation emitted from an electric spark and had the same numerical value as the ratio between the electrostatic and electromagnetic units of charge. It may be convenient to summarize these early results which are still of scientific interest although the values obtained are now of little importance in view of the more accurate measurements made subsequently. The results obtained by using light are given in Table I, those obtained using electromagnetic radiation in Table II and those from the ratio of the units in Table III. They have been adjusted to values in a vacuum for uniformity.

TABLE I

Velocity of light in vacuo

Date	Author	Method	Result (km/sec)	Limits of error
1676	Roemer	Jupiters satellites	214,000	
1726	Bradley	Aberration of stars	301,000	
1849	Fizeau	Toothed wheel	315,000	
1862	Foucault	Deflection of light by rotating mirror	298,000	± 500
1872	Cornu	Toothed wheel	298,500	± 900
1874	Cornu	Deflection of light	300,400	± 800
1878	Michelson	Deflection of light	300,140	± 700
1879	Michelson	Deflection of light	299,910	± 50
1882	Newcomb	Deflection of light	299,810	± 30
1882	Michelson	Deflection of light	299,853	± 60
1908	Perrotin & Prim	Toothed wheel	299,901	± 84

TABLE II

Velocity of electromagnetic radiation in vacuo

Date	Author	Method	Result (km/sec)
1891	Blondlot	Lecher wires	297,600
1895	Trowbridge & Duane	Lecher wires	300,300
1897	Saunders	Lecher wires	299,700
1899	MacLean	Free space	299,100

TABLE III

Ratio of electromagnetic to electrostatic units

Date	Author	Value (km/sec) in vacuo
1857	Weber and Kohlrausch	310,800
1868	Maxwell	284,300
1869	Thomson and King	280,900
1874	McKichan	289,700
1879	Ayrton and Perry	296,100
1880	Shida	295,600
1883	Thomson, J. J.	296,400
1884	Klemencic	302,000
1888	Himstedt	301,000
1889	Thomson, W.	300,500
1889	Rosa	300,090
1890	Thomson, J. J. and Searle	299,690
1891	Pellat	301,010
1892	Abraham	299,220
1897	Hurmuzescu	300,190
1898	Perot and Fabry	299,870
1899	Lodge and Glazebrook	301,000

CHAPTER 2
The Nature of Light
1. THE CORPUSCULAR AND WAVE THEORIES

Since the seventeenth century scientists have been much occupied with the nature of light, and the rival claims of the corpuscular theory (according to which light consists of small particles) and of the wave theory (according to which it consists of a wave motion propagated in space or in the ether) have at times been advanced with considerable heat. Such a heated controversy was both unnecessary and unfortunate and can only have arisen from a mistaken idea of the role of theory in the development of science. It is always helpful to have a physical picture of a new phenomenon or in other words it is helpful to compare it with phenomena which have already been studied. A good deal is known about the behaviour of particles and if light consists of particles this previous knowledge can be utilized, and in the same way the existing knowledge about wave propagation can be used if light is a form of wave motion. However it is axiomatic that a new phenomenon will differ in some respects from any that has been investigated previously and that the physical picture can therefore never be complete. Early theories are simply an aid to further investigations and should never be taken too seriously or held too dogmatically. This was the attitude taken by Newton although the fact that he tended to favour the corpuscular theory is said to have delayed the development and acceptance of the wave theory for many years. Newton's first scientific paper on the optical spectrum was strongly criticized by Hooke. In reply Newton protested that his views on colour were in no way bound up with any particular conception of the ultimate nature of optical processes. However, in order to relate them to Hooke's hypothesis, and having made sure that colour is an inherent characteristic of light, he inferred that it must be associated with some definite quality of the corpuscles or aether vibrations. The corpuscles corresponding to different colours would, he remarked, like sonorous bodies of different pitch, excite vibrations of different types in the aether. With remarkable insight at various places in his "Opticks", (1931) Newton refers to this double aspect of light, but after his discovery of the inverse square law of gravitation he began to regard the aether hypothesis as superfluous. In the eighteenth century the wave theory of Huygens was developed further particularly by Fresnel and Young and was used to explain interference and diffraction phenomena.

2. THE NATURE OF LIGHT

A great advance in our understanding of light followed from the work of Faraday and Maxwell in which it was deduced that light is one form of electromagnetic radiation, and the extension of Maxwell's theory by Lorentz to the electron was fruitful in explaining the dispersion and absorption of light and various magneto-optic effects. In spite of these successes the wave theory failed to explain many experimental results. The quantum theory according to which light can be emitted or absorbed in quanta of energy equal to hf, where h is Planck's constant, was developed and provided a quantitative explanation of these results.

The experiments devised for measuring the velocity of light are for the most part fairly straightforward measurements of the time of travel of a pulse of light or radio waves over a measured distance, and a theory of propagation is needed only for calculating small corrections which need to be made. In the measurements made with radio waves the dimensions of the waves become comparable with those of the apparatus and in these circumstances the corrections become of greater significance. In all cases the wave theory of light provides the simplest way of dealing with the problems and evaluating the corrections.

2. THE WAVE THEORY IN OPTICS

According to the wave theory energy is transmitted from a source to a receiver not by means of corpuscles of matter but by the motion of some medium. Sound is transmitted by air and in this case the air particles move backwards and forwards in the direction of transmission. A signal caused by dropping a stone in water is transmitted by the particles of water which, however, move up and down at right angles to the direction of propagation of the signal. The transmission of light is considered to be somewhat similar, being propagated in a medium, the nature of which cannot be specified, moving at right angles to the direction of motion. A wave of this form is represented by an equation similar to equation (1.5) which will be rewritten

$$I = a \cos 2\pi \left(\frac{t}{\Gamma} - \frac{x}{\lambda} \right) = a \cos \frac{2\pi}{\lambda}(vt - x) \qquad (2.1)$$

where a, Γ and λ are constants and x is the distance from a source. At a fixed distance I varies between values of $\pm a$ which is called the amplitude of the wave, the change of t for a complete cycle being Γ which is called the periodic time. The frequency $f = 1/\Gamma$. At a fixed time I varies between the values $\pm a$ as x is increased the distance between successive peaks being λ, the wave length. The velocity of the wave is clearly the product of the frequency and the wavelength.

Some properties of waves, used in velocity measurements, are readily illustrated from equation (2.1). The resultant displacement I of two waves is obtained by adding the displacements of the individual waves. If the waves differ only in phase the value is

$$I = a \cos 2\pi \left(\frac{t}{T} - \frac{x}{\lambda}\right) + a \cos 2\pi \left(\frac{t}{T} - \frac{x + \delta x}{\lambda}\right)$$

$$= 2a \cos \pi \frac{\delta x}{\lambda} \cos 2\pi \left(\frac{t}{T} - \frac{x + \delta x/2}{\lambda}\right) \quad (2.2)$$

and represents a wave of the same wave length and frequency but with an amplitude $2a \cos(\pi \delta x/\lambda)$ which varies between the limits of $\pm 2a$ according

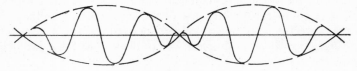

Fig 2.1

to the value of δx. The two waves thus interfere and annul each other when $\delta x = \lambda/2$.

The formula for group velocity, which is important when light extending over a band of wave lengths is used in velocity measurements, is obtained by combining two waves of slightly different wave length travelling at different velocities. The resultant wave is given by

$$a \left[\cos \frac{2\pi}{\lambda}(x - vt) + \cos \frac{2\pi}{\lambda + \delta\lambda}\{x - (v + \delta v)t\}\right]$$

which neglecting small terms reduces to

$$2a \cos \frac{2\pi}{\lambda}(x - vt) \cos \pi \left(x \frac{\delta\lambda}{\lambda^2} - t \frac{v\delta\lambda - \lambda\delta v}{\lambda^2}\right). \quad (2.3)$$

The envelope of the wave is shown in Fig. 2.1 and although the velocity of the individual waves is still v that of the group is

$$\frac{v\delta\lambda - \lambda\delta v}{\delta\lambda} = v - \frac{\lambda\delta v}{\delta\lambda}. \quad (2.4)$$

In some determinations it is the group velocity which is measured. It differs from the phase velocity v when the velocity changes with wave length, which is the case in any medium except a vacuum.

3. THE ELECTROMAGNETIC THEORY OF LIGHT

It was found experimentally that the force between charges and magnetic poles depends on the medium and equations (1.1) and (1.2) should therefore be replaced by

$$F = \frac{q_1 q_2}{\varepsilon d^2}, \quad F = \frac{m_1 m_2}{\mu d^2} \tag{2.5}$$

where ε is called the permittivity and μ the permeability of the medium. The constants K_1 and K_2 have been dispensed with as their values can be incorporated in ε and μ. Since the forces still exist when the medium is a vacuum, empty space can be regarded as having electrical and magnetic properties with values of permittivity ε_0 and permeability μ_0. The values for other media are usually expressed relative to those for a vacuum so that in general $\varepsilon = \varepsilon_0 \varepsilon_r$ and $\mu = \mu_0 \mu_r$. Moreover in order to simplify electromagnetic equations it is usual to include 4π in the force equations which become

$$F = \frac{q_1 q_2}{4\pi\varepsilon_0 \varepsilon_r d^2} \qquad F = \frac{m_1 m_2}{4\pi\mu_0 \mu_r d^2} \tag{2.6}$$

The forces cannot easily be measured directly but some of the electrical quantities can be determined with great accuracy. A capacitance for example can be measured in different ways so as to give the value $\varepsilon_0 \mu_0$, which is thus an experimentally determined quantity, and can be regarded as a fundamental property of space. The value found is $1/c_0^2$, and when the quantity $1/\sqrt{\varepsilon_0 \mu_0}$ appears in electromagnetic equations, as it often does, it is usually replaced by c_0. In electrostatic and magnetic problems μ_0 and ε_0 appear separately but the relationship can be maintained if an arbitrary value is given to only one of them. In the MKS system of units the value $4\pi \times 10^{-7}$ henry/m is ascribed to μ_0, and ε_0 is then $1/c_0^2 \mu_0$ or $8 \cdot 854 \times 10^{-12}$ farad/m. When these values are put in equation (2.6) and $\mu_r = \varepsilon_r = 1$ for a vacuum, the equations serve to define the units of charge and magnetic pole and the complication of two systems of units is avoided.

The concept of electric and magnetic fields was introduced by Faraday to explain the forces exerted by charged and magnetized bodies on distant bodies. The intensity E of the electric force on a small test charge q is given by

$$F = qE \tag{2.7}$$

and from equation (2.6) it is seen that a single point charge gives rise to a field

$$E = q/4\pi\varepsilon r^2. \tag{2.8}$$

In a similar way the intensity of magnetic field due to a point magnetic charge is given by

$$H = m/4\pi\mu r^2. \tag{2.9}$$

The experimental laws of electromagnetism were incorporated by Maxwell into his well known field equations which in cartesian coordinates take the following form in which σ is the conductivity of the medium and space-charges are absent.

$$\frac{\partial E_x}{\partial z} - \frac{\partial E_z}{\partial x} = -\mu \frac{\partial H_y}{\partial t}$$

$$\frac{\partial E_z}{\partial y} - \frac{\partial E_y}{\partial z} = -\mu \frac{\partial H_x}{\partial t} \tag{2.10}$$

$$\frac{\partial E_y}{\partial x} - \frac{\partial E_x}{\partial y} = -\mu \frac{\partial H_z}{\partial t}$$

$$\frac{\partial H_z}{\partial y} - \frac{\partial H_y}{\partial z} = \sigma E_x + \varepsilon \frac{\partial E_x}{\partial t}$$

$$\frac{\partial H_x}{\partial z} - \frac{\partial H_z}{\partial x} = \sigma E_y + \varepsilon \frac{\partial E_y}{\partial t} \tag{2.11}$$

$$\frac{\partial H_y}{\partial x} - \frac{\partial H_x}{\partial y} = \sigma E_z + \varepsilon \frac{\partial E_z}{\partial t}$$

$$\frac{\partial}{\partial x}(\mu H_x) + \frac{\partial}{\partial y}(\mu H_y) + \frac{\partial}{\partial z}(\mu H_z) = 0 \tag{2.12}$$

$$\frac{\partial}{\partial x}(\varepsilon E_x) + \frac{\partial}{\partial y}(\varepsilon E_y) + \frac{\partial}{\partial z}(\varepsilon E_z) = 0. \tag{2.13}$$

There are an enormous number of solutions to these equations and it is their generality which enables them to be applied to a vast range of problems including the propagation of waves under widely different experimental conditions. For the case of an insulating medium for which $\sigma = 0$ it is possible to obtain a set of equations such as

$$\frac{\partial^2 H_x}{\partial x^2} + \frac{\partial^2 H_x}{\partial y^2} + \frac{\partial^2 H_x}{\partial z^2} = \mu\varepsilon \frac{\partial^2 H_x}{\partial t^2} \tag{2.14}$$

which represents a wave travelling with the velocity $1/\sqrt{\mu\varepsilon}$. If we introduce the restriction that the wave travels in the z direction, a particular solution of this equation is

$$H_x = A \cos \frac{2\pi}{\lambda}(z - vt) \qquad (2.15)$$

which is the equation of a wave moving in the z direction with a velocity $v = 1/\sqrt{\mu\varepsilon}$. A more rigorous analysis of this case shows that a magnetic wave must always be accompanied by an electric wave and that the electric and magnetic fields are at right angles to each other and to the direction of propagation of the wave. The wave therefore might be represented by fig. 2.2. The amplitudes of the electric and magnetic components are related by $\sqrt{\varepsilon E} = \sqrt{\mu H}$ so that $E/H = \sqrt{\mu/\varepsilon}$ or in a vacuum $\sqrt{\mu_0/\varepsilon_0}$. This quantity, like $1/\sqrt{\mu_0 \varepsilon_0}$, is a fundamental constant of space and since it has the dimensions of an impedance it is called the wave impedance.

4. THE EXPERIMENTS OF HERTZ

The electromagnetic waves predicted by Maxwell's theory were first produced in practice by Hertz (1887). He employed a spark discharge because the radio valve had not then been invented, but this crude method had the advantage of giving rather short waves, about 2m long. He was thus able to carry out some very important experiments showing the similarity between radio and light waves. His source consisted of two brass plates to each of which was connected a heavy wire terminated in a small sphere; the two spheres being slightly apart. The wires were connected to an induction coil which produced opposite charges on the two plates, hence acting as a capacitor. When the charge was sufficiently high the air insulation between the spheres broke down and there was an oscillatory discharge between them lasting for a few cycles. The process was repeated each time the primary circuit of the induction coil was broken, and a succession of pulses of high frequency waves was thus produced. The receiver was a loop of wire, each end being terminated in a small sphere, the size of the loop being such that the circuit was in tune with the source frequency. When the receiver was placed in the field of the source sparks were produced across the spheres, and the distance between the spheres across which the sparks could be obtained served as a rough measure of the intensity of the field.

With this simple equipment Hertz was able to show that the electric and magnetic intensities are at right angles to each other and to the direction of propagation. He produced a system of standing waves by adjusting the distance of a reflecting sheet in front of the source and also demonstrated that the waves were refracted through a prism and were polarized.

5. THE DIFFRACTION OF LIGHT

A plane wave of light does not cast a perfectly sharp shadow of an object, but some light extends within the geometrical shadow and there are bands of varying brightness immediately outside it. This diffraction of light was studied by Grimaldi in the middle of the seventeenth century and later by Hooke, Newton and Young, but the first satisfactory explanation was given by Fresnel. He considered that the wave front could be divided into zones

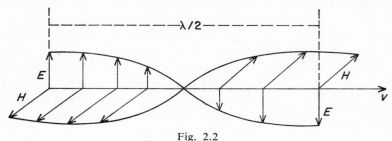

Fig. 2.2

such that successive zones differ in their distance from the object by $\lambda/2$. They arrive at the object in opposite phase and if only two zones were concerned their effects would cancel. The resultant effect at any point is obtained by integrating the effects of all parts of the wave front and the integrals are now known as Fresnel's integrals. In any experiment involving slits, screens, obstructions, or the transmitting and receiving aerials used in radio work, diffraction effects are present and their importance depends on the relative dimensions of the apparatus and the wave length of the radiation. For example very long radio waves—the longest used for radio communication are about 20 km long—are little affected by obstacles such as hills and buildings and extend round them even more readily than do sound waves. The short radio waves used for television programmes on the other hand are readily obstructed and demand something approaching a "line of sight" path.

6. THE NATURE OF LIGHT

All the experimental evidence obtained so far confirms that waves extending from low radio frequencies, through the infrared, optical, ultra-violet, X-ray and γ-ray bands are of the same basic nature. The long radio waves can be examined in detail and the fields associated with them can be measured by magnetic and electrical instruments. It can be regarded as an experimentally established fact that a plane electromagnetic wave, including light, is basically of the form shown in Fig. 2.2. The propagation of the wave is affected by the objects in its path and depends on the relative size of the objects compared with the wave length. The detailed calculations of the path become very complex even when the physical details of the objects can be defined, and become insoluble when they cannot be defined. This is the condition when radiation interacts with atomic electrons and it is not to be expected that the wave theory can deal with these problems. The wave theory, however, is adequate to deal with any problems associated with the measurement of the velocity of light.

CHAPTER 3

Standards and Accuracy of Measurement

The determination of velocity consists in the measurement of a time and a distance, and the final accuracy depends on the accuracy of these two measurements, together with the precision with which the time can be set to correspond to a fixed distance or *vice versa*. In most of the determinations which have been made in the past the limit has been set by the precision of this setting but in recent experiments the precision has been greatly improved and the errors in the time and length measurements have assumed greater importance. The length in particular must now be measured with an accuracy approaching that of the definitive standard of length itself. It is important therefore to consider these standards.

1. DEFINITIVE STANDARD OF LENGTH

Standards of length existed in the earliest civilizations and at the time of Edward I of England, for example, the Greek foot (12.45 in) Roman foot (11·65 in) and the Saxon foot (13·2 in) were all in use. That far sighted monarch attempted to secure some uniformity and the Ulna or yard was established by an act of 1305. Although defined in terms of barley corns it consisted of a brass rod which is still in existence and said to be 0·04 in short of the present yard. New yard standards were constructed from time to time but the rise of engineering science required a length standard which could be used with higher precision. Two bar standards marked by fine lines to define the yard were made by George Graham in 1742 for the Royal Society. They were sent to Paris and one was returned, inscribed with the French standard of length, the half-toise. Some idea of the precision with which these bars could be used can be obtained from a paper by Evelyn read at the Royal Society in 1798. He found that the difference between Graham's bar and one he had had made himself was 0·00130 in and that the distance between the half-toise marks said by the French Academy to be 38·355 in was 38·356 in. It seems therefore that the bar standards were defined to about 2 parts in 10^5 at that time. In 1834 the copies in the Houses of Parliament were destroyed by fire and 40 new standard yards were constructed, two of them being sent to the USA where they served as the national standards for a number of years.

3. STANDARDS AND ACCURACY OF MEASUREMENT

There had for long been a feeling that it would be desirable to base the length standards on some natural phenomenon and many experiments were made to define the length of a bar which would have a time of swing of exactly 1 sec when acting as a pendulum. This method was not very satisfactory, however, because apart from the experimental difficulties of supporting the bar, the time of swing depends on the gravitational force which varies significantly over the earth's surface. The timing of the swing of a pendulum at different places is in fact now the usual way of studying the variations in gravity.

In France ancther method was attempted. In 1790 the Academy of Sciences was instructed to make proposals for a system of weights and measures based on natural constants. It recommended that the standard of length should be one ten millionth part of the distance between the North Pole and the equator measured along the meridian passing through the Paris Observatory. The unit was called the metre but the measurement in terms of the meridian was so difficult and inaccurate that in effect the metre was defined as the length of the bronze bar which had been constructed as a provisional standard. In 1798 when the geodetic measurements had been completed a new metre was constructed and made the legal standard. The metre was thus based on the length of the earth's meridian although it became in practice simply the length of a metal bar.

The international standardization of units was first placed on a formal footing at the Metric Convention in Paris in 1875, when the International Bureau of Weights and Measures at Sèvres was set up. In 1889 a new International Prototype Metre was made of an alloy of 90% platinum and 10% iridium. National copies could be compared with the original with a precision of 1 part in 10^7. The search for an invariable natural standard was not abandoned however. Babinet had suggested as early as 1827 that the wavelength of a monochromatic source of light might be used in this way and it became a practical possibility when towards the end of the nineteenth century Michelson and Benoit measured the length of the metre in terms of the cadmium red line. As time went on it was increasingly used as a standard because the technical applications of interferometry provided the most convenient way of calibrating the gauges used in the engineering industry. It was not formally adopted as a standard, however, because owing to a number of factors which perturbed the natural wavelength emitted by the atom it would not have enabled the accuracy of the definition to be increased. The construction of light sources using single isotopes of metals greatly reduced the band of wavelengths radiated and also enabled interferometer techniques to be used over greater lengths. In consequence in 1960 the metre was defined as equal to 1,650,763·73 vacuum wavelengths of the orange line (spectroscopic designation $2p_{10} - 5d_5$) emitted by the krypton atom of mass 86, and experience showed that it could be used to an accuracy of 1 part in 10^8. This is the accuracy of length measurement that can be achieved at the present time (1968). It is possible that lasers will eventually provide a higher accuracy but the wavelength of the radiation which they emit is strongly influenced by the geometry of the resonator and they must therefore be calibrated in terms of the krypton source.

This brief historical survey shows that the accuracy of the standards of length has always been higher than the setting accuracy in the measurement of the velocity of light.

2. DEFINITIVE STANDARD OF TIME

The standard of time until 1955 was the rotation of the earth on its axis. Successive transits of a particular star across the meridian of a place give the time interval of one sidereal day and this is the smallest unit of time that can be defined by astronomical measurements. Clocks serve the purposes of subdividing this interval into 86,400 equal seconds and enable the unit to be used outside the observatories. It is an interesting fact that the name of Galileo is associated with both the first telescopes and the first reasonably accurate clocks based on a pendulum. The type of telescope used for time determination is known as a transit instrument and the first of these was made by the Danish astronomer Roemer, who was also the first to obtain a value for the velocity of light. An improved instrument installed at Greenwich Observatory by Halley in 1772 enabled transit times to be noted to 1 sec, and a later instrument installed by Airy in 1851 gave a result accurate to $\pm$ 20 msec for an observation of a group of stars. A new form of telescope known as the photographic zenith tube was built at the U.S. Naval Observatory in 1911. It was based on an instrument previously made by Airy but embodied many improvements, which increased its accuracy for time measurements to about ± 5 msec. With some further modifications this instrument was gradually adopted by most observatories.

The times of star transit are recorded or photographed and used together with time markers from the standard clocks to determine the "rates" of the clocks. The clocks must of course run uniformly between the observations if they are to divide the day into equal seconds and this can be checked by comparing a number of clocks which together constitute the working standard. Early pendulum clocks such as those of Tompion and Graham (1722) kept time to about 1 sec/day. Improvements made by Airy in 1872 increased the accuracy to 0·1 sec/day and modern pendulums such as that due to Shortt keep time with a uniformity of about 0·02 sec/day, which is comparable to the accuracy of the individual astronomical observations. The next great advance in time keeping arose from the need to measure the frequencies of radio waves. For this purpose tuning fork and quartz frequency standards were developed. A typical modern quartz standard operates at a frequency of 5 Mc/s. The frequency is divided to give a decade scale of frequencies or time intervals and finally drives a clock dial. Quartz clocks such as the Essen-ring and the Giebe and Scheibe bars not only provided the necessary short time intervals but were much better timekeepers over long periods than pendulum clocks. They enabled the errors of the astronomical measurements to be smoothed out and provided time uniform to 0·001 second per day, or 1 part in 10^8. It was then found that the rotation of the earth itself varied by several parts in 10^8 and thus limited the accuracy of timekeeping by astronomical means. This limitation has been removed by the introduction of atomic clocks the first of which was made by Essen and Parry (1958). Atomic clocks provide a unit of time accurate to 1 part in 10^{11} and the second of time is now defined as the duration of 9,192,631,770 periods of the radiation corresponding to the transition between the two hyperfine levels of the ground state of the atom of caesium 133. As with the unit of length, errors in the unit of time have never yet presented a limitation in the accuracy of measuring the velocity of light.

3. APPLICATION OF THE DEFINITIVE STANDARDS

Up to 1960 the definitive standard of length was made available by the distribution of copies of the standard yard or metre to the National Laboratories where they were used to calibrate working standards such as engraved scales and tapes. In this way it was possible to refer back to the standard although each stage of comparison introduces some additional errors. The krypton lamp now constitutes an independent standard.

Geodetic base-lines such as those used in velocity determinations have been measured by invar tapes or wires 24m or 50m long and suspended in catenary under a definite tension. The accuracies quoted are often only the precision of repetition and may not include the errors of the tape calibration.

Calibrated tapes are no longer sufficiently accurate for the most accurate velocity measurements, and the distances must be measured as directly as possible in terms of wave lengths by interferometer techniques. Before the days of radio time signals time measurements involved the setting up of a pendulum and timing it by reference to astronomical measurements, but radio signals now make accurate time available throughout the world. The signals can be used to check a local standard, like a quartz clock, and now that time signals are controlled by atomic clocks the full accuracy to 1 part in 10^{11} is available by fairly simple means.

4. THE SETTING ACCURACY

In general, the velocity of light measurements are made by adjusting either a frequency or length to some criterion such as the maximum or minimum signal on a detecting instrument. Either the frequency or length usually remains constant during the measurement and can be determined at leisure using the appropriate techniques and care to give the required precision. The other quantity must be measured in a short time during which the indicating instrument is maintained at its maximum or minimum value. If this measurement is repeated in a short time during which other pameters may be assumed to remain constant the scatter of the measurements may be regarded as the error of setting. It includes the errors of judgment in setting to the adopted criterion and the error in measuring the length or frequency when it has been set.

It is clearly impossible to discover any errors introduced at any part of the experiment which are less than the setting error. The experiment should therefore be designed with the object of making this as small as possible. The effects of altering the experimental conditions can then be found and errors which would otherwise remain as hidden systematic errors can be eliminated.

5. SYSTEMATIC ERRORS

A study of the experimental results obtained in the past shows that their accuracy has been limited by systematic errors rather than by precision of setting. A consideration of possible systematic errors is therefore a most important part of the experimenter's task when the main object of the work, as in the velocity of light measurements, is to obtain an accurate quantitative result. Errors in the values of the definitive standards are outside his control but other systematic errors can arise due to the fact that the chosen criterion, such as the minimum of a signal, does not in fact represent the setting giving the true relationship between distance and time. This is the kind of error which can be studied when the precision of setting is high enough. Another error could arise in transforming the measured value of velocity to the velocity in vacuo. In experiments not carried out in a vacuum the correction depends on the refractive index of the light used at the atmospheric conditions prevailing at the time of the experiment. The correction for refractive index is of importance to the majority of velocity of light determinations and it will therefore be treated in some detail in the next section.

6. THE REFRACTIVE INDEX OF AIR

The refractive index n of a medium was defined in the development of optics as the ratio of the velocity of light in a vacuum to that in the medium. In electromagnetic theory this ratio is $\sqrt{\mu_0 \mu_r \varepsilon_0 \varepsilon_r}/\sqrt{\mu_0 \varepsilon_0}$. We have therefore

$$n = \sqrt{\mu_r \varepsilon_r}. \tag{3.1}$$

3. STANDARDS AND ACCURACY OF MEASUREMENT

The refractive index of air varies with its constitution, its temperature and pressure, and the frequency of the radio or light waves. The composition of the air remains very uniform except for the amount of water vapour and it can therefore be regarded as consisting of two components, dry air and water vapour. The variation with frequency is due to the atomic and molecular resonances which occur throughout the spectrum. The value may therefore be expected to hold for all frequencies below 40 GHz, where the strong oxygen absorption at 60 GHz may begin to have an effect. Water vapour has a weak absorption at 22 GHz but according to Van Vleck's (1942) theoretical work this should have a negligible effect on the refractive index. The water vapour absorption in the region of 300 GHz which is over 1000 times stronger than that at 22 GHz becomes noticeable at frequencies above 100 GHz. As a result of these strong absorptions in the microwave and infrared regions the refractive index of water vapour is greatly reduced and in the optical region it does not differ greatly from that of dry air while in the radio frequency region it is 20 times as great. The water vapour correction is therefore particularly important in the radio frequency measurements. In the optical region air and water vapour are free from resonant absorption and although the effects of resonances in the infrared and ultraviolet regions are present, they can be taken into account by use of a relatively simple formula. The refractive index of dry air is usually given for the standard conditions of 0°C and 760 mm Hg and the value at other conditions must be computed. Although air approximately obeys the ideal gas laws, a more accurate empirical formula has been given by Barrell and Sears (1939). This can be written

$$(n_{t,p} - 1) = \frac{(n_{0,760} - 1)p}{760.606(1 + 0.003661\,t)} [1 + (1.049 - 0.0157\,t)\,10^{-6}p] \tag{3.2}$$

where n is the refractive index, p the pressure in mm of Hg and t the temperature in °C. This formula applies to air free from carbon dioxide. The carbon dioxide content is only 0·03% by volume in the "open air" and it is adequate to treat it as a perfect gas, so that

$$(n_{t,p} - 1) = (n_{0,760} - 1) \frac{p}{760} \cdot \frac{273}{T} \tag{3.3}$$

where T is the absolute temperature. Water vapour is a polar gas with an electric dipole moment and therefore obeys Debye's (1929) equation of the form

$$\varepsilon - 1 = p' \left(\frac{A}{T} + \frac{B}{T^2} \right) \tag{3.4}$$

where p' is the pressure that the vapour exerts assuming that the ideal gas laws are obeyed at any fixed temperature. Barrell and Sears (1939) found values for p' and A as follows

$$p' = p(1 + 2.4 \times 10^{-5} p) \qquad (3.5)$$

$$A = 1.725 \times 10^{-4}. \qquad (3.6)$$

Essen and Froome (1951) found from microwave measurements and the above value for A that

$$B = 0.9913 \qquad (3.7)$$

giving as a resultant formula for water vapour

$$(n-1) 10^6 = \frac{86.24}{T} p \left(1 + \frac{5748}{T}\right)(1 + 2.4 \times 10^{-5} p). \qquad (3.8)$$

Essen and Froome also made very careful measurements of the refractive index of dry air and the constituents of dry air at frequencies between 9 GHz and 72 GHz and using their values in equations (3.2), (3.3) and (3.8) they obtained the following extrapolation formula for moist air

$$(n_{tp} - 1) 10^6 = \frac{0.37884 p_1}{1 + 0.003661 t} \left[1 + (1.049 - 0.0157 t) 10^{-6} p_1 + \right.$$
$$\left. + \frac{1.774 p_2}{273 + t} + \frac{86.24 p_3}{273 + t} \left(1 + \frac{5748}{273 + t}\right)(1 + 2.4 \times 10^{-5} p_3)\right] \qquad (3.9)$$

where p_1 is the partial pressure of dry air free from carbon dioxide, p_2 is the partial pressure of carbon dioxide, p_3 is the partial pressure of water vapour and $p = p_1 + p_2 + p_3$. All pressures are in mm of Hg at 0° C and at standard gravity. This equation can be used for all atmospheric conditions with temperatures between —20° C and +60° C and unsaturated vapour pressures less than about 100 mmHg without introducing additional errors in n as great as that of the actual measured values which was ± 1 part in 10^7 and at all frequencies below 30 GHz. Results obtained for dry air at the National Bureau of Standards are in agreement with the NPL results within 1 part in 10^7. This full accuracy is required for the most precise measurements of the velocity of light but it can be simplified for many applications by assuming that dry air behaves as an ideal gas and that water vapour behaves as an ideal gas at any one temperature. The simplified formula then becomes

$$(n_{tp} - 1) 10^6 = \frac{103.49 p_1}{T} + \frac{177.4 p_2}{T} + \frac{86.26}{T}\left(1 + \frac{5748}{T}\right) p_3 \qquad (3.10)$$

where $T = 273 + t$ (273 was used in the development of the formula for simplicity and also because of the uncertainty in the precise value of the absolute zero temperature). Equation (3.10) is still fully accurate at the temperature and pressure of the measurements (approximately 20° C and 760 mmHg) but introduces an error of 5×10^{-7} at the extremes of the above range of temperature and with normal water vapour pressures. For most purposes the effect of the carbon dioxide can be neglected and p_2 can then be put equal to 0 and $p = p_1 + p_3$. The small amount of carbon dioxide present is thus assumed to have the same refractive index as dry air. A set of tables based on equation (3.10) has been prepared by the NPL and it has also been used for the preparation of a chart calculator which is available commercially.

The refractive index of air in the optical region can be calculated from a formula derived by Edlén (1953). This is

$$(n_s - 1) \, 10^8 = 6432 \cdot 8 + 2949810/(146 - \sigma^2) + 25540/(41 - \sigma^2) \quad (3.11)$$

where n_s represents the refractive index of "standard" air at 15° C, 760 mm Hg pressure (corrected), and containing 0·03 % by volume of carbon dioxide, and σ is the vacuum wave number in μ^{-1}, i.e. the number of vacuum wavelengths in 1 micron. With little loss in accuracy this can be expressed in the Cauchy equation

$$(n_s - 1) \, 10^8 = 27259.9 + 153.58/\lambda_{\text{vac}}^2 + 1.318/\lambda_{\text{vac}}^4 . \quad (3.12)$$

For air at t° C and total pressure p in mm Hg and containing water vapour at p_3 mm Hg the refractive index, as given by Barrell and Sears (1939), is

$$(n_{t,p} - 1) = \{0.001387188(n_s - 1) \, p \, (1 + \beta_t \, p)/(1 + \alpha t)\} - $$
$$- \{(a - b/\lambda_{\text{vac}}^2) \, p_3/(1 + \alpha t)\} \quad (3.13)$$

where

$$\alpha = 0.003661 , \qquad \beta_t = (1.049 - 0.0157 \, t) \, 10^{-6}$$
$$a = 6.24 \times 10^{-8} , \quad b = 0.0680 \times 10^{-8} .$$

Modulated light (or for that matter, radio and microwave radiation) travels through a dispersive medium with a group velocity u, which is related to the phase velocity v by the formula

$$u = v - \lambda \frac{dv}{d\lambda} . \quad (3.14)$$

For transmission through air we can express this velocity in terms of a "group" refractive index, n_g, where

$$n_g = n - \lambda \frac{dn}{d\lambda} . \quad (3.15)$$

Here n is the true refractive index.

For visible and near infra-red "light" the Cauchy form of Edlén's equation (3.12) is accurate and the most convenient for differentiation. Thus the "group" refractive index for standard air becomes

$$(n_{gs} - 1)\,10^8 = 27260 + 460.8/\lambda^2 + 6.60/\lambda^4. \qquad (3.16)$$

In particular, this is essentially the form of equation used by Bergstrand (see Chapter 9) in his velocity determinations using air paths of the order of 10 km.

7. PRESENTATION OF THE RESULTS

In order that experimental measurements of the same quantity can be compared it would clearly be an advantage if they could be presented in some uniform way particularly as regards the limits of error. A study of the problem shows unfortunately that this aim cannot be realized. The accuracy is usually limited by systematic errors, whose presence may be suspected but which cannot be studied experimentally. The reliability of the given limits of error therefore depends on the consideration given by the experimenter to possible sources of systematic error, and the more careful he is the larger will be the limits. The only part of the experimental work that lends itself to statistical treatment is the repetition of a number of readings made under what are believed to be the same conditions. If the mean of such a set of n observations of x is $\bar{x}$ then the standard deviation of a single observation is

$$\sigma = [\Sigma(\bar{x} - x)^2/(n - 1)]^{\frac{1}{2}}. \qquad (3.17)$$

It is often found that σ does not differ much from the average value of $(\bar{x} - x)$ or the deviations of single measurements from the mean. It does not decrease significantly as n is increased beyond a fairly small value, say 20. Limits are sometimes given as the standard deviation of the mean which may be called σ_m and is obtained from

$$\sigma_m = [\Sigma(\bar{x} - x)^2/(n - 1)\,n]^{\frac{1}{2}}. \qquad (3.18)$$

This quantity gradually decreases as n is increased and it must be used with considerable care. Indeed it should be used only when it can be checked experimentally by taking a number of sets of observations and finding that σ for these sets is in fact equal to the value of σ_m obtained for one set. Whether σ or σ_m is used the limits should never be reduced below a value corresponding to the reading accuracy of any instrument used in the measurement; otherwise the use of the above formulae could lead to a situation where, to give a simple example, a temperature accurate to $\pm 0\cdot001°$ C could be measured with a thermometer which could be read to only $0\cdot1°$ C.

If sensibly used σ and σ_m provide a useful and well understood way of describing the precision or repeatability of a set of measurements, but even so they do not necessarily make the results of different experiments strictly comparable. It often happens that the experimeter knows that some systematic errors will be reduced by varying the experimental conditions, for example by reversing the direction of rotation of a mirror. One complete measurement is then obtained by averaging the results obtained under these different conditions, so that a certain amount of averaging has already taken place, and the value of σ will be less than if the measurement consisted of a single observation.

Since it is not possible to present results in a uniform manner they can only be judged by a critical examination of the work as a whole and of the author's treatment of the errors.

Many reviews have been published of earlier measurements of the velocity of light and the reviewers have recorded what they regard as the most likely value. One of the most thorough of these is that of Dorsey (1944), but although he studied the papers in great detail and pointed out how they were unsatisfactory in many respects, his final value was incorrect and his limits were too small. Another review carried out by Birge also led to incorrect conclusions. These reviews show how difficult it is for anyone, and particularly anyone who has not himself performed similar experiments, to judge the work of others. The simple average of all the results of optical experiments having a scatter between measurements not exceeding ± 250 km/sec would have given a value of 299,800 km/sec, and of those having a scatter not exceeding ± 75 km/sec a value of 299,792 km/sec. Although the closeness of the last result with the present accepted value of 299,792·5 km/sec must be fortuitous, simple averaging certainly gave a better value than the 299,773 km/sec $\pm$ 10 km/sec given by Dorsey or the 299,776 km/sec $\pm$ 4 km/sec given by Birge. The careful work of the reviewers thus gave both an incorrect value and incorrect limits of error. In our present work we shall give the results as published together with the scatter of individual observations where this is available.

CHAPTER 4

Optical Methods 1908—1940

During the period 1908—1940 a number of important measurements were made and although they gave only a modest increase in the accuracy they introduced new techniques which eventually led to a notable improvement. The first two measurements to be discussed were carried out by Michelson (1927) and by Michelson, Pease and Pearson (1935). Michelson followed a suggestion made earlier by Newcomb and used a rotating mirror with a number of faces. The speed of rotation of the mirror was adjusted so that light reflected from one face was returned to the neighbouring face. There was still a small displacement of the image, which was measured but the time of travel of the light was almost equal to that taken by the mirror to rotate through $1/n$ revolutions where n is the number of sides on the mirror. The accuracy of setting depended on the precision of observing the position of the image as in Foucault's original method, but in general fewer errors are introduced in setting to a zero or near zero value than in reading a deflection. Systematic errors in particular are likely to be less.

In his final determination made in 1935 in collaboration with Pease and Pearson, Michelson used a similar multi-sided mirror but confined the light path to a mile-long evacuated pipe in order to eliminate the effect of the refractive index of the atmosphere. At the time this was quite an ambitious engineering achievement but it is doubtful whether it improved the accuracy of the result. The refractive index correction was not by any means the main source of error. A more fundamental step towards improving the accuracy had already been taken by Karolus and Mittelstaedt (1928).

If we consider the original Fizeau experiment in which the light was chopped by means of a toothed wheel it is clear that the precision of setting the frequency of rotation of the wheel can be expressed as a fraction of a tooth. If the frequency or the distance is increased so that the returning light hits not the next tooth but the nth then the precision of measurement, other things being equal, is n times as great. There are limitations to increasing the distance because of the reduction in light intensity and Karolus therefore considered ways of increasing the frequency. He abandoned mechanical methods and modulated the intensity of the light by means of the electro-optic effect of a Kerr cell. In this way he was able to use a modulation frequency of the order of 10 MHz compared with the 10 kHz used by Michelson and co-workers. The measurements were, however, made over a correspondingly shorter distance of 40 m compared with 16 km and in consequence the accuracy was about the same. It seems to have been intended to increase the distance to 3 km but the method was not used with longer distances until the later work of Bergstrand. Similar techniques with comparatively short distances were used by Anderson (1937, 1941) and by Hüttel (1940) and similar accuracies were obtained.

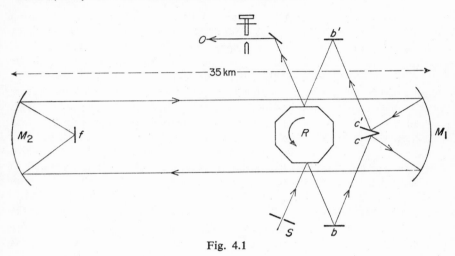

Fig. 4.1

1. MICHELSON'S DETERMINATION IN 1924—1926

The apparatus was arranged symmetrically as shown in Fig. 4.1 and such an arrangement always has the advantage of eliminating some possible systematic errors. Light from a Sperry arc behind the slit S was reflected at one face of the rotating prism R and at the plane mirrors b and c to the concave mirror M_1 which produced a parallel beam. This was transmitted a distance of 35 km (from Mount Wilson to Mount San Antonio) to the mirror M_2. The return beam was reflected from M_1 to the plane mirrors c' and b' to the opposite face of the rotating prism and to the eye piece at O. The position of the image was determined by a micrometer. The rotating mirror was driven by an air blast at such a speed that it rotated by one face during the travel of the light so that the image was almost at the position it occupied when the mirror was at rest. The image position was observed first with the mirror rotating in one direction and then at the same speed in the reverse direction. If the two small deflections are α_1 and α_2 the double deflection of the return image was

$$2\phi = \pi - (\alpha_1 - \alpha_2) = 16\pi nD/c$$

where D is the length of the light path, n the number of rotations/sec and c the velocity of light.

Thence
$$c = 16nD/(1 - \beta)$$
where
$$\beta = (\alpha_1 - \alpha_2)/\pi.$$

The angle $(\alpha_1 - \alpha_2)$ was of the order of 0·002. The distance was measured by the United States Coast and Geodetic Survey with an estimated accuracy of one part in 2 million and the speed of rotation was determined by a stroboscopic method. A small mirror attached to the prong of an electrically driven tuning fork reflected light from the revolving mirror. The speed was controlled by means of a counterblast so as to obtain a stationary picture for a few seconds during which the image position was measured. This speed was 530 turns/sec, giving for the eight-sided mirror a modulation frequency of 4240 Hz. The frequency of the fork was 132·25 Hz, i.e. 1/32 of the modulation frequency. It was measured in terms of a pendulum which in turn was compared with a standard pendulum. A preliminary result of $c_o = 299{,}820$ km/sec was obtained, and was thought to be accurate to about 30 km/sec. The correction applied for the refractive index of the air was too large and Michelson later changed the value to 299,802 km/sec. A further series of measurements was made in 1925 using essentially the same equipment with an improved timing mechanism. The tuning fork was valve driven and was compared directly with a standard pendulum. An extensive set of definitive measurements using 12-sided and 16-sided mirrors was carried out during 1926 and the final value was given as $299{,}796 \pm 4$ km/sec. The spread of the individual results, each being the average of a number of settings, was about 60 km/sec and no attempt was made to investigate systematic errors. The limit given must therefore be regarded as optimistic. Nevertheless the internal consistency of the results suggests that this was Michelson's best determination and, as we now know, the result is very close to the value accepted at the present time.

2. MICHELSON, PEASE AND PEARSON 1935

This final measurement was proposed and planned by Michelson who however, died in 1931 when only 36 of the 233 series of observations had been made. It was sponsored by the University of Chicago, the Mount Wilson Observatory, the Rockefeller Foundation and the Carnegie Corporation. A level site was selected on the Irvine Ranch near St. Anto, and the U.S. Coast and Geodetic Survey established a new base line and measured it on three occasions. The greater part of the light path was inside a pipe 1·6 km long and 1 m in diameter made from Armco-iron steel sheets, 0·7 m wide rolled and formed into sections 18 m long which were joined by elaborate seals. The pressure in the pipe was maintained between 0·5 and 5·5 mm Hg according to the leaks. The optical system, though basically the same as that used before was modified to direct the light into the pipe and to give a light path of either 12·8 km or 16 km by multiple reflections. The scheme is shown in Fig. 4.2. When the mirror was not rotating the outgoing and return-

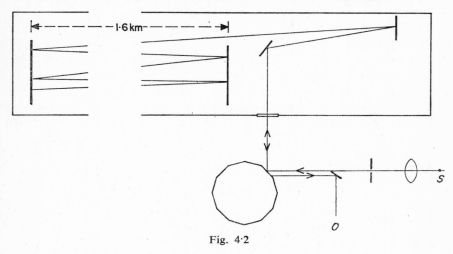

Fig. 4·2

ing beams were reflected at the same face and not from opposite faces as previously. The speed of rotation was measured as before by the stroboscopic pattern of light reflected from the rotating system and the prong of a tuning fork. The frequency of the fork was then determined stroboscopically by reference to a pendulum clock. The rate of the pendulum was determined in terms of a chronometer which in turn was rated by means of time signals. The process was thus rather complex and must have introduced some errors; but it was the opinion of the authors that these could not account for the discrepant results which were obtained. Full details of the results are given by the authors. There were 233 series of observations each series consisting of about 10 individual measurements. The spread of the mean values for the series is still nearly 100 km/sec although most of the results fall within ± 25 km/sec of the final average value. The authors point out that although the mean value is given as $299{,}774 \pm 11$ km/sec the averages of different groups varied by an amount which could not be explained.

3. KAROLUS AND MITTELSTAEDT 1928

Two ways of using the Kerr cell as a light shutter were discussed by Karolus and Mittelstaedt. The cell consists of two electrodes immersed in a liquid such as nitrobenzene. When a high voltage is applied to the electrodes a plane polarized beam of light entering the cell is converted into an elliptically polarized one because the doubly refracting cell produces a phase difference between the two components of the beam, which are respectively parallel and perpendicular to the electric field. The phase lag is proportional to the square of the field intensity and may be represented by

$$\phi = 2\pi B l E^2 \qquad (4.1)$$

where B is the Kerr constant, l is the length of the electrodes in the direction of the beam and E is the electric intensity between them. In one scheme the

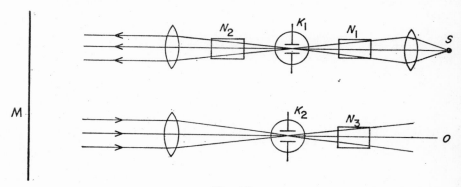

Fig. 4.3

cell is preceded by a polarizing Nicol prism N_1 and followed by an analysing Nicol N_2 (see Fig. 4.3) and the incident beam is polarized at 45° to the direction of the field. The intensity of the emergent light is given by

$$I = I_0 \sin^2 \phi/2. \qquad (4.2)$$

It is usual to apply a steady voltage as well as an alternating voltage to the cell so that it operates at a suitable, nearly linear, part of its voltage characteristic, as shown in Fig. 4.4. The returning beam passes through a second light shutter consisting of a Kerr cell K_2 operated exactly in phase with K_1 and a Nicol N_3. The amount of light passing through this shutter depends on the phase of the returning beam and will have a minimum value for a certain value of frequency, just as in the original Fizeau experiment. Preliminary measurements using the first order light minimum gave an accuracy of 1% but it was decided to alter the optical arrangement to reduce the dependency on the intensity of the light source. The second Nicol N_2 was eliminated as shown in Fig. 4.5 and the light returning to K_2 was then still elliptically polarized. K_2 was rotated through 90° relative to K_1 so as to produce an opposite polarization and thus convert the light back to a plane polarized beam. The Nicol N_3 is crossed relative to N_1 and no light reaches the observer when the frequency is such that the path is a whole number of wavelengths.

The oscillator producing the voltage applied to the Kerr cell should be a powerful one so that the gap between the electrodes can be large to let

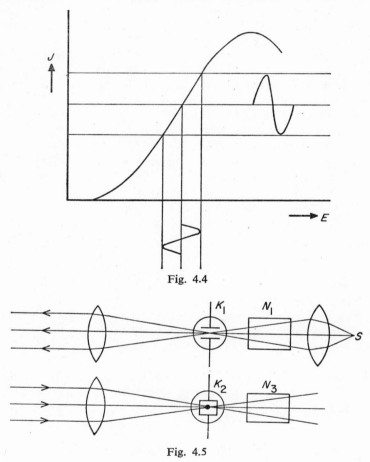

Fig. 4.4

Fig. 4.5

through a reasonable amount of light. In practice a gap of 2·5 mm was used with an alternating voltage of 2000 V on which was superimposed a steady voltage of 6000 V. Frequencies of 3·5 MHz to 7·2 MHz were used with two path lengths near 250 m and 330 m, the order of the light minima being 3, 4, 5 and 8.

The frequency was measured by comparing it with an overtone of a tuning fork having a frequency of 2506·5 Hz or 2511·5 Hz and measured with an accuracy of 1 part in 10^5 in terms of a time standard.

The experiment was carried out in a corridor 40 m long and the long light path achieved by multiple reflections. The distance was measured by means of a steel wire under a tension of 20 kg, and the wire was calibrated in a separate measurement in terms of length standards. Altogether 755 measurements were made giving the result 299,778 km/sec $\pm$ 20 km/sec; the spread of the individual measurements was $\pm$ 240 km/sec and the mean error of a single observation was $\pm$ 100 km/sec. The spread of the means of 5 series of measurements under different conditions of frequency, length, or order number was 31 km/sec, and it is interesting to note that the set which the authors, from their measurements, considered to be the most accurate gave a result the furthest away from what is now regarded as the true value. Corrections were applied for the refractive index of air and for the transmission through glass and the nitrobenzene of the Kerr cell but no mention is made of a correction for group velocity. Since a modulated light beam was used it is the group velocity that was measured and a correction of about 6 km/sec should be applied (Andersen, 1941) increasing their value to 299,784 km/sec.

4. ANDERSON 1937—1941

Anderson used a Kerr cell to modulate the light but made two important changes in technique. The receiving light shutter was not another Kerr cell but a photoelectric cell with the consequences that the electrical capacitance in the Kerr cell circuit was reduced permitting the use of higher frequencies and it was moreover no longer necessary to keep the two cells balanced. Another change which may well be more important was the introduction of two light paths, one short and one long, as shown in Fig. 4.6. The position of mirror M_3 was adjusted to give a minimum intensity of the signal from the photocell. Assuming that the photoelectric cell is linear and that the intensities of the light from the two paths are the same the voltages developed by the cell are

$$e_1 = E_0 \sin \omega t \; ; \quad e_2 = E_0 \sin (\omega t + \phi)$$

where $\phi = S\omega/c$ is the phase difference. The resultant voltage

$$e_1 + e_2 = E_0 [\sin \omega t + \sin (\omega t + \phi)] = 2E_0 \cos \phi/2 \sin (\omega t + \phi/2) .$$

This is a minimum when

$$\phi(= S\omega/c) = n\pi$$

or

$$c = \omega S/n\pi = 2fS/n$$

where $n = 1, 2, 3, \omega = 2\pi f$, f is the frequency of the radio frequency voltage and S is the optical path difference. The important thing to notice is that the minimum does not depend on the intensity of the source, and this remains true if the intensities of the two beams are unequal. The frequency was a harmonic of a quartz oscillator and was therefore nominally constant and the path length was changed to obtain the minimum light intensity.

The signal received from the photocell is at the modulation frequency. It can be either rectified to operate a meter or an audible beat-note can be obtained with another oscillator having nearly the same frequency. The

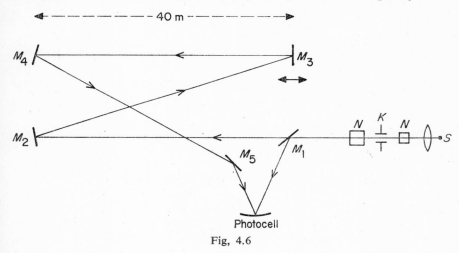

Fig, 4.6

amplitude of the signal is indicated by the amplitude of the audio note and the ear can be used as a sensitive detector for reducing it to zero or a minimum value. Anderson also used a third method of detection. Instead of applying a steady voltage to the Kerr cell he applied a voltage alternating at the mains power frequency (60 Hz). The signal obtained is then modulated at the mains frequency and can be detected as an audible note without the use of an additional beat oscillator.

The distance was measured by a stretched invar steel tape checked by a standard 2 m bar, and two measurements made at an interval of six months agreed to 5 parts in 10^6. The frequency was measured by comparison with harmonics of a 10 kHz standard in a radio receiver the resultant beat note being measured by an audio frequency oscillator. The audio note was measured to about 10 Hz in a radio frequency of several million Hz but there was some additional uncertainty because of the frequency drift of the quartz oscillators with temperature. The result was called preliminary but 651 observations were made giving a value of 299,764 ± 12 km/sec. The spread of the individual readings was 120 km/sec and the difference between two groups of over 100 observations was 16 km/sec. Anderson later applied a correction for group velocity bringing the value to 299,771 km/sec.

For the final measurements a number of improvements were made to the equipment. The optical system was rearranged as shown in Fig. 4.7 to facilitate the measurement of the path difference. Light from the source S was divided by the half silvered mirror M_6 one part being reflected from M_3 to the photocell and the other part along the long path M_1, $M_7 M_5$ and

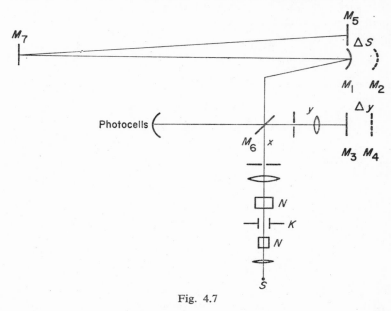

Fig. 4.7

back again to the photocell. The combined light passes through a short focus lens to give practically a point image in the cell. The mirror M_3 which is on a lathe-bed is adjusted to give a minimum signal. The path difference is then an odd number of half wave lengths and

$$2S + 2x - 2y = (2n + 1)/\lambda/2$$

where λ is the wavelength of the radio frequency modulation, S is the long path, x is the short additional path between M_6 and M_1 and y is the path between M_6 and M_3. M_1 and M_2 are then rotated so that M_1 is ineffective and M_2 reflects the light directly back to M_6. M_3 is adjusted to the position M_4 to give a new minimum. Then

$$2x + 2\Delta S - 2y - 2\Delta y = \lambda/2$$

and

$$2S - 2\Delta S + 2\Delta y = n\lambda.$$

It is therefore unnecessary to measure x and y. A new source of light was used consisting of a 1000 watt water cooled mercury arc, a new type of Kerr cell was designed and in the detector side of the circuit an eleven stage electron multiplier tube was added. The Kerr cell was operated with a biasing voltage of between 5000 and 10,000 V and a modulating voltage of about 1000 V at 19·2 MHz. The frequency was derived from the 50 kHz frequency standard and the path length was about 170 m.

Nearly 3000 measurements were made giving a result of 299,776 ± 14 km/sec. The total spread of the results is not given but the means of groups of over 100 measurements differ by 60 km/sec indicating the presence of systematic errors. Anderson mentions that one such error may arise from the two beams striking different parts of the cathode surface of the photocell.

5. HÜTTEL 1940

Hüttel's method is very similar to Anderson's. He used a photocell detector, the difference between two light paths as the effective distance, and a source which was modulated at a low frequency as well as at a radio frequency. Most of the measurements were made with the optical arrangement shown in Fig. 4.8. Light from the source is chopped at a frequency of 200

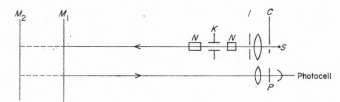

Fig. 4.8

Hz by the mechanically driven disc C, passes through a lens and an adjustable iris I to the Nicol—Kerr cell light shutter, and is then reflected from either M_1 or M_2 to the photocell. The purpose of the mechanical chopper was to give an audio signal output which could be amplified more readily than a d.c. signal. Hüttel considered the errors very carefully including that due to the time-lag in the photocell. To overcome this he introduced a finely ground plate P in front of the photocell so as to diffuse the light, although there was in consequence a considerable loss of signal. He also observed the signal at a point where it was changing most rapidly instead of at a flat minimum. The intensity of the light in front of the light shutter varies roughly as a sine curve as shown in Fig. 4.9 and Hüttel pointed out that it would be preferable to make measurements at points such as A and B rather than at the maxima or minima positions. M_1 corresponds to point A

and M_2 to point B. It was of course necessary to allow for the diminution of signal with increasing distance and the iris was adjusted so that the intensities were equal when no r.f. modulation was applied. The position of mirror M_1 was set to give a convenient signal and then M_2 was set to give

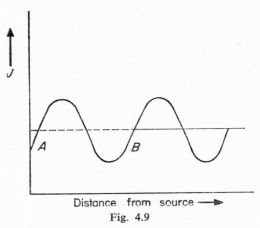

Fig. 4.9

the same signal, the iris being used to compensate for the light loss. Hüttel made measurements at three different frequencies, near 5, 8 and 12 MH_z and at three different orders. The total number of results recorded was 135 and the weighted average was 299,768 ± 10 km, the spread of groups of about 20 results being 35 km/sec. The fractional error due to the length measurement is estimated as $\pm 4 \times 10^{-5}$ and that due to frequency as $\pm 7 \times 10^{-5}$. Some part of these rather large errors was probably systematic and could easily account for the large error in the result.

CHAPTER 5

Electrical Methods 1900 — 1940

In chapter 1 two different methods of determining c by electrical measurements were briefly described. Weber and Kohlrauch measured the charge on a capacitor in electrostatic and electromagnetic units and Blondlot measured the wavelength and frequency of a radio wave transmitted along a pair of parallel wires. Both of these methods were improved and great care was taken with the calculation of systematic errors. Presumably it was not appreciated at the time that the refractive index of air for radio waves depended greatly on the humidity and no records appear to have been taken of the precise atmospheric conditions. Even so the results were probably as reliable as those obtained by optical methods.

1. ROSA AND DORSEY (1907)

A number of capacitors in the form of concentric spheres, concentric cylinders and plates were constructed and their dimensions were accurately determined by metrological methods and also by gravimetric methods, in which the weight of water which they held or displaced was measured. The capacitance between two concentric spherical conductors is simply

$$C = \varepsilon r_1 r_2/(r_1 - r_2) \tag{5.1}$$

where r_1, r_2, are the radii, but corrections had to be made for the hole, through which the inner sphere was suspended by a silk thread, and for the charging wires. The corrections were only a few parts in 10^5 and the calculated capacitor was believed to be known to 2 parts in 10^5 giving an error of 1 part in 10^5 in velocity. The value of r_1 was approximately 12·67 cm and two innier balls of 10·11 cm and 8·87 cm were used. The capacitances were approxmately 45·2 $\mu\mu F$ and 26·6 $\mu\mu F$ respectively. The capacitance of the oncentric cylinders is

$$C = \varepsilon(l + \delta l)/(2 \log_e r_1/r_2) \tag{5.2}$$

where l is the length and as before r_1 and r_2 are the two radii. δl is an end correction which is made small by means of guard cylinders, and can be calculated and also checked experimentally. The capacitor could be used without guard cylinders and δl could then be eliminated by varying the length l. The radii were approximately 7·25 cm and 6·25 cm making the capacitance per unit length about 38 $\mu\mu$F. The capacitance between parallel circular plates is

$$C = \varepsilon a/4\pi d = \varepsilon(r + \delta r)^2/4d \qquad (5.3)$$

where a is the area, d the distance between the plates and r the radius. δr is a term due to the effect of the gap between the plate and the guard ring. The distance between the plates was varied between 1·6 mm and 9·6 mm.

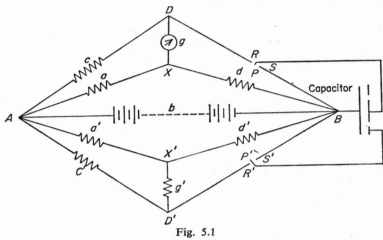

Fig. 5.1

The error of the calculated value was probably considerably greater than that of the other types of capacitor largely because of an error due to any lack of parallelism.

In all cases the capacitance can be expressed as

$$C = \varepsilon A \qquad (5.4)$$

where A is calculated from the dimension of the capacitor.

The capacitance was then measured in terms of a resistance by means of a Maxwell bridge and also by a differential galvanometer method. The bridge adapted for use with a capacitance having a guard ring is shown in Fig. 5.1. The corresponding arms a, a', c, c', etc., have the same values of resistance and the potentials D, D' and x, x' are therefore the same. PRS and $P'R'S'$ are joined to two sets of brushes on a commutator which charge and discharge the capacitor. The circuit ensures that the guard ring is charged to the same potential as the main condenser. From bridge network theory it can be shown that

$$C = \frac{a[1 - a^2/(a + c + g)(a + b + d)]}{f_1 cd[1 + ab/c(a + b + d)][1 + ag/d(a + c + g)]} = B/Rf_1 \qquad (5.5)$$

where f_1 is the number of times/sec the capacitor is charged, B is a quantity containing simply the measured values of the various resistors and R is the value of the unit of resistance in terms of the units of length and time and the magnetic constant. The unit of resistance is determined at the National Laboratories in terms of a calculated inductance. One method is that due to Lorenz in which a metal disc is rotated in a magnetic field and the potential between the rim and the centre is balanced against that across a resistor. The field is that between two coils of calculable mutual inductance with respect to the disc and carrying the same current as that in the resistor. The value of the resistor can then be expressed as

$$R = \mu f_2 M \qquad (5.6)$$

where M is a function of the dimensions of the coils, f_2 is the frequency of rotation of the disc and μ is the permeability of the medium. From equations (5.4), (5.5) and (5.6) we have

$$C = \varepsilon A = B/\mu f_2 M f_1 \qquad (5.7)$$

from which

$$\mu \varepsilon = B/A M f_1 f_2 . \qquad (5.8)$$

Rosa and Dorsey used the accepted value of the unit of resistance.

In the differential method the charging or discharging current passed through one coil of a differential galvanometer while a steady current shunt-

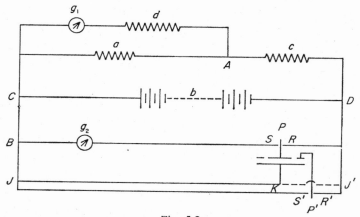

Fig. 5.2

ed off from a portion of a high resistance passed through the other. The circuit is shown in Fig. 5.2. The continuous lines show the connections for sending the discharging current through the galvanometer. For measuring the charging current JK is disconnected, $J'K$ connected, and the connections at S, R and also at S', R' are interchanged. If E is the e.m.f. of the battery the current through g_2 is f_1CE where f_1 is the number of charges/sec. The current through g_1 is E $a/cd(1+A_2)$ where

$$A_2 = g_1/d + a(1/d + 1/c + g_1/cd). \tag{5.9}$$

In the condition of balance when the current through g_1 equals that through g_2

$$f_1 C = a/cd(1 + A_2) \tag{5.10}$$

where A_2 is a relatively small term. As before

$$C = B/Rf_1 \tag{5.11}$$

where B is a function of the measured resistances.

The commutator for charging and discharging the capacitor was driven by a small motor generator fitted with heavy fly-wheels and run from a large storage battery. The speed was regulated by a slight friction applied to the rim of one of the fly wheels, so that under favourable conditions it was maintained constant to a few parts in 10^5. The speed was timed by means of chronometers which were rated against a Riefler pendulum clock which was itself compared with standard time signals. Frequencies between 250 Hz and 500 Hz were used.

A great many measurements were made and the results were reduced to the standard condition of 20° C. The humidity was kept low mainly because of leakage problems but the effect of the presence of water vapour on the dielectric constant does not seem to have been considered. The dielectric constant at 20° C and 760 mm Hg pressure was assumed to be 1·00055. This would correspond to extremely dry conditions and a more usual figure would be 1·00065.

The results are given for the average of small groups of several measurements and summarized in Table I.

TABLE I

Capacitor	No. of measurements	c Weighted mean km/sec	Spread of values for small groups km/sec	Average deviation km/sec
Spherical 1	64	299,621	80	18
Spherical 2	81	299,659	70	16
	147	299,637	60	11
Cylindrical	329	299,609	130	27
with guard	174	299,626	150	30
Plate 1	29	299,626	440	
Plate 2	45	299,630	310	

The final value is based only on the spherical and cyclindrical capacitors and is given as

$$c = 299{,}630 \pm 30 \text{ km/sec}$$

or in a vacuum

$$c_0 = 299{,}710 \pm 30 \text{ km/sec}$$

It is now believed that the value of the unit of resistance at that time was 520 parts in 10^6 too large. If allowance is made for this the value of c_0 becomes 299,788 km/sec $\pm$ 30 km/sec and if as appears likely an additional correction is needed for the humidity of air the result would become 299,803 km/sec $\pm$ 30 km/sec. Since there is no definite evidence about this the value will be taken as 299,788 km/sec.

2. MERCIER 1923

The method used by Mercier (1923, 1924) exploits his work on the measurement of high frequencies. He had shown that the frequency of one oscillator could be set to a harmonic of that of another oscillator by listening to the beat note between them and setting this to zero. By a series of stages it was in this way possible to set the frequency of an oscillator operating at hundreds of megahertz to a multiple of a low frequency oscillator. The low frequency oscillator itself was measured in terms of a standard pendulum. Very high frequencies could thus be measured to 1 part in 10^5. The process was simplified by synchronizing, or locking, some of the stages to a harmonic of the preceding stage. Before this time the values of very high frequencies had been deduced from the values of wavelength and an assumed value for velocity. The wavelength was often found by measuring the distance between successive nodes on a pair of parallel wires (Lecher wires), or by detecting the resonance of a circuit, the natural resonant frequency of which could be calculated. For example the frequency of a lossless circuit consisting of a capacitance C and inductance L is $1/2\,\pi\sqrt{LC}$. Mercier gave a full historical introduction to this type of experiment and pointed out that although Maxwell's theory predicts that the velocity along a lossless pair of wires should be the same as that in air, an extension of the theory by Sommerfeld and Mie established that it would be influenced by the diameter and spacing of the wires and on the frequency used. Experimental attempts to check the theory were not very successful. Mercier starts therefore with a theoretical treatment and by making suitable approximations he arrived at the simple relationship

$$\frac{v_0 - v}{v_0} = \frac{\delta}{4r \log_e D/r} \tag{5.12}$$

where v_0 is the velocity in air, v that along the wires, D is the spacing between the wires, r is their radius and δ is a quantity known as the skin depth of penetration at the frequency concerned. The full theory will not be given here but a consideration from circuit theory shows the nature of the problem. The velocity along the wires can be shown to be $1/\sqrt{LC}$ where L and C are the inductance and capacitance of the wires per unit length. The capacitance does not change with frequency but the inductance does. At very high frequencies the current flows in a very thin surface layer of the wires which act simply as a "guide" for the electromagnetic wave; but at low frequencies the current penetrates the wires which act more as conductors in the usual sense. The penetration can be found from Maxwell's equations for a plane wave in a conducting medium. The expressions σE must be retained in equations (2.11) and the wave equations then take the form

$$\frac{\partial^2 E_x}{\partial z^2} = \frac{\mu \varepsilon \delta^2 E_x}{\partial t^2} + \mu \sigma \frac{\partial E_x}{\partial t}. \qquad (5.13)$$

Similar equations hold for H and the equations represent a damped wave motion. If, following Bleaney and Bleaney (1957), consideration is limited to a plane wave travelling parallel to the x-axis, then E_x and H_x become zero and a plane transverse wave is obtained. If this is plane polarized with a single component of E parallel to the y-axis the field equations become

$$\partial E_y/\partial x = -\mu(\partial H_z/\partial t)$$
$$-\partial H_z/\partial x = \varepsilon(\partial E_y/\partial t) + \sigma E_y. \qquad (5.14)$$

It will be found that the velocity now depends on frequency but if consideration is restricted to a single frequency it can be assumed that E_y and H_z vary with time as a complex exponential

$$\exp j\omega(t - nx/c_0)$$

where n is the complex refractive index. The equations then become

$$j\omega(n/c_0) E_y = j\omega\mu H_z$$
$$j\omega(n/c_0) H_z = (j\omega\varepsilon + \sigma) E_y. \qquad (5.15)$$

Elimination of E_y or H_z shows that the equations are satisfied if

$$n^2 = \mu_r \varepsilon_r - j(\sigma\mu_r/\omega\varepsilon_0) \qquad (5.16)$$

where μ has been replaced by $\mu_r\mu_0$, ε by $\varepsilon_r\varepsilon_0$ and c_0 by $1/\mu_0\varepsilon_0$. Replacing n by n-jk and separating the real and imaginary parts gives

$$n^2 - k^2 = \mu_r \varepsilon_r$$
$$nk = \sigma\mu_r/2\omega\varepsilon_0 \qquad (5.17)$$

and the wave is propagated as

$$\exp\{j\omega(t - nx/c_0)\} = \exp(-\omega kx/c_0)\exp\{j\omega(t - nx/c_0)\} \quad (5.18)$$

showing that the amplitude of the wave decreases according to the value of k. In the case of a good conductor the value of $j\omega\varepsilon$ in the second equation (5.15) is very small compared with σ. The value of ε and of ε_r in equation (5.17) can be taken as zero making

$$n = k = (\sigma\mu_r/2\omega\varepsilon_0)^{\frac{1}{2}}. \quad (5.19)$$

If $n = k$ is put equal to $c_0/\omega\delta$ expression (5.18) for the propagation of the field components of the wave becomes

$$\exp(-x/\delta)\exp j(\omega t - x/\delta). \quad (5.20)$$

δ has the dimensions of a length and it is called the "skin depth". The amplitude of the wave falls to $1/e$ of its initial value in the distance δ. The value of δ in terms of conductivity, frequency and permeability is

$$\delta = (\sigma\omega\mu/2)^{-\frac{1}{2}} = (\pi\sigma f\mu)^{-\frac{1}{2}}. \quad (5.21)$$

For the case of copper σ 0·64 ohm^{-1}m^{-1} $\mu_r = 1$ and $\mu_0 = 4\pi \times 10^{-7}$ making $\delta = 6\cdot2f^{-\frac{1}{2}}$ cm. For a frequency of 80 MHz used by Mercier $\delta = 1\cdot1 \times 10^{-4}$ cm.

The way in which the skin penetration affects the inductance of a pair of parallel wires cannot be expressed very simply. Hartshorn (1947) for example expresses the inductance of a length l of two wires of radius r and spacing D as

$$L = 4 \times 10^{-7} l[\log_e(D/r) - D/l + \mu_r(1 - U)/4] \quad (5.22)$$

where U is a tabulated quantity depending on the skin depth. At high frequencies $(1-U)$ has the value $2\delta/r$. Remembering that the velocity varies as $1/\sqrt{L}$ and neglecting small quantities the effect on velocity reduces to the expression (5.12) given by Mercier. It should be remembered that although the correction term is large the calculations are quite rigorous since only those quantities have been neglected which are known by calculation to be negligibly small. As Mercier points out the wire is not perfect, but later work at higher frequencies shows that such imperfections are not important when the skin depth is as great as 10^{-4} cm.

The experimental arrangement (see Fig. 5.3) consisted of two wires 11 m long stretched horizontally 2 m above the ground and 2 m from the walls and ceiling of the room. The oscillator was inductively coupled to one end of the wires which were connected at this point by a metal bridge B. Another bridge C could be moved along the wires forming a closed circuit which could be set to resonance with the oscillator. The movable bridge was carefully designed so as to make good electrical contact and to remain at right angles across the wires. Resonance was observed by the current flowing through a lead sulphide crystal rectifier D connected across the far end of the wires. The current was measured by means of a microammeter or galvanometer and the bridge could be set to give a maximum current with a precision of better than 0·1 mm. The distances were measured on an

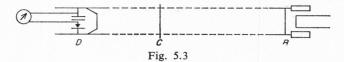

Fig. 5.3

invar tape stretched at about 1 m below the wires and graduated in mm at the regions near 0, 2 m and 4 m. The value of v was obtained from the measured frequency and wavelength.

The precision of measurement soon revealed that the results were affected by surrounding bodies and by movements of the observer. A very large room was needed so that the wires could be situated further away from any other equipment, but since this was not available conditions were chosen empirically which gave the most consistent results. A set of 12 measurements made under the same conditions gave a total spread of 54 km/sec and results were then obtained with a number of different wires and spacings. For various reasons it was decided to reject the results obtained with the thickest wire and smallest spacings leaving those recorded in Table II. The corrections are given in the table because one of the author's objects was to check their validity.

TABLE II

Diameter of wire mm	Separation of wire mm	Measured velocity km/sec	Calculated correction	Corrected velocity c km/sec
1·16	79·2	299,466	260	299,726
	39·1	299,417	303	299,720
	19·3	299,342	366	299,708
2·06	80·3	299,568	130	299,698
	40·1	299,515	155	299,670

Mean 299,700 km/sec estimated uncertainty $\pm$ 30 km/sec

No details are given of the atmospheric conditions but if we apply the correction of 95 km/sec which is appropriate for average laboratory conditions, the result would become

$$c_0 = 299,795 \pm 30 \text{ km/sec}$$

5. ELECTRICAL METHODS 1900—1940

Mercier points out that the value is about 150 km/sec lower than the commonly accepted value at that time, and suggests several possible sources of error in his own measurements but does not suggest that the optical value might be wrong.

3. SUMMARY OF VALUES OBTAINED BETWEEN 1905 and 1940

It might be convenient at this stage to summarize the values obtained during this period to supplement those given in tables I, II and III in Chapter 1. None of these results are any longer of interest from the point of view of obtaining the best value for the constant.

TABLE III

Values obtained for the velocity of light in vacuo during the period 1905—1940

Year	Author	Method	Spread of values km/sec	Limits of error km/sec	Value of velocity km/sec
		Electrical methods			
1907	Rosa and Dorsey	Ratio of units	100	±30	299,788
1923	Mercier	Lecher wires	50	±30	299,795
		Optical Methods			
1924	Michelson	Rotating mirror		±30	299,802
1926	Michelson	Rotating mirror	60	± 4	299,796
1928	Karolus and Mittelstaedt	Kerr cell	240	±20	299,778
1935	Michelson, Pease and Pearson	Rotating mirror	100	±11	299,774
1937	Anderson	Kerr cell	120	±12	299,771
1940	Hüttel	Kerr cell	35	±10	299,768
1941	Anderson	Kerr cell	60	±14	299,776

CHAPTER 6

Cavity Resonator Measurements

Mercier's measurement of the velocity of light followed as an application of his work on the measurement of high radio frequencies In the same way the cavity resonator method was the direct outcome of the extension of frequency measuring techniques to the microwave region of the spectrum. Such waves travel down a hollow metal cylinder and if the cylinder is closed at both ends and is exactly a whole number of half-wavelengths long, resonance is established. The wavelength of the standing waves in the cylinder differs from that in free space and at 10 GHz for example ($\lambda = 3$ cm) resonance is obtained in a cylinder about 2·5 cm long and 2·5 cm in diameter. If one of the end-plates is attached to the head of an ordinary micrometer, a wide range of resonant frequencies is obtained with a simple robust and accurate instrument. The resonant frequencies can be calculated from the dimensions and the velocity of light, but in practice it is much easier and more accurate to calibrate the scale of the instrument by direct reference to a standard oscillator. Essen and Gordon—Smith (1945) were among the first to develop a heterodyne method of doing this. In the course of such calibrations at the N.P.L. Essen (1946) measured the dimensions of a number of commercial cavity wavemeters and compared the theoretical and measured values of resonant frequency. The results can be interpreted as a measurement of the velocity of light, the value obtained being 299,770 ± 30 km/sec. The values were corrected for the effect of the resistance of the walls as calculated from theory but in view of the lack of mechanical precision no great importance was attached to these initial results. It is however interesting to note that the experimental scatter was considerably less than that of any earlier measurement.

He concluded that measurements made in a vacuum with a carefully constructed cavity resonator should give a more accurate value for the velocity than that given by the methods previously employed.

1. THEORY OF THE CAVITY RESONATOR

Electromagnetic theory was applied to cavity resonators by J. J. Thomson as long ago as in 1893 and then more thoroughly by Lord Rayleigh in 1897. It was surely one of the most remarkable successes of any scientific theory to be able to apply the laws of electromagnetism, established from experiments with inductance coils, resistors and capacitors, to frequencies which are so high that these components of equipment no longer have any separate significance. At the frequencies of hundreds of megahertz used by Mercier it is still possible to consider inductance, resistance and capacitance as being distributed along a pair of wires but in the microwave region these concepts are no longer useful. The conductors and dielectrics simply place boundary conditions on the electromagnetic wave and restrict the propagation to certain modes with different configurations of the electric and magnetic fields. The fields, introduced by Faraday as an aid to the understanding of electric and magnetic phenomena, play an increasingly important part. They are almost like physical entities and provide a model which is essential for the design of equipment.

The mathematical analysis of cavity resonators is carried out by solving the general field equations with the appropriate boundary conditions. It can be effected exactly, in terms of known mathematical functions, only for certain simple shapes such as the rectangular prism, the cylinder and the sphere. For cylindrical wave guides it is convenient to express the field equations in circular cylindrical coordinates r, ϕ, x which are related to the axes of the cartesian coordinate system by

$$x = x, \quad y = r \sin \phi, \quad z = r \cos \phi.$$

Considering for example the vector H we have in place of H_x, H_y and H_z

$$H_r = H_y \sin \phi + H_z \cos \phi$$

$$H_\phi = H_y \cos \phi - H_z \sin \phi$$

$$H_x = H_x.$$

The wave equations in cylindrical coordinates are of the form

$$\frac{\partial^2 H'_x}{\partial r^2} + \frac{1}{r} \frac{\partial H'_x}{\partial r} + \frac{1}{r^2} \frac{\partial^2 H'_x}{\partial \phi^2} = [j\omega\mu(\sigma + j\omega\varepsilon) - \gamma^2] H'_x \qquad (6.1)$$

where it is assumed that intensities involve time in the form $e^{j\omega t}$ and distance in the form $e^{-\gamma x}$, where γ is the complex propagation constant $\gamma = \alpha + j\beta$. In a dielectric $\sigma \ll \omega\varepsilon$ and the equation becomes

$$\frac{\partial^2 H'_x}{\partial r^2} + \frac{1}{r} \frac{\partial H'_x}{\partial r} + \frac{1}{r^2} \frac{\partial^2 H'_x}{\partial \phi^2} = -(\gamma^2 + \omega^2 \mu\varepsilon) H'_x \qquad (6.2)$$

2. FREQUENCIES OF CAVITY RESONATORS

The solution of equation (6.2) is obtained in terms of Bessel functions. The equations for a cavity resonator are then obtained by summing waves of equal intensity travelling in opposite directions in a cylinder closed at both ends. Resonance occurs when there are a whole number, including zero, of half or whole waves of the electric field pattern in the coordinate directions and the mode of resonance is conveniently designated by the three suffixes l, m, n where

- l = number of full-period variations of the radial component around the angular coordinate;
- m = number of half-period variations of the angular component along the radial coordinate;
- n = number of half period variations of the radial component along the axial coordinate.

The frequencies of all the modes in a lossless guide can be expressed by the equation

$$(f_{lmn} D)^2 = c^2 \left[\left(\frac{\Gamma_{lm}}{\pi} \right)^2 + \left(\frac{nD}{2L} \right)^2 \right] \quad (6.3)$$

where D, L are respectively the diameter and length of the resonator in cm, n is the number of half waves contained in the length of the resonator, Γ_{lm} is the mth root of $J_l(x) = 0$ for those modes having no magnetic field parallel to the x axis (*TM* or *E* modes) and the mth root of $J_l'(x) = 0$ for those modes having no electric field parallel to the x axis (*TE* or *H* modes). $J_l(x)$ is a Bessel function of the first kind of order 1 and $J_l'(x)$ is its first derivative. $c = c_0/\sqrt{\varepsilon_r}$ is the velocity of propagation in the medium of relative permittivity ε_r. Values of Γ_{lm} for a few low order modes are given in Table I.

TABLE I

Values of Γ for low order modes

Mode designation	Γ
TE_{11}	1·84118378
TM_{01}	2·40482556
TE_{21}	3·05423693
TM_{11}	3·83170597
TE_{01}	3·83170597
TE_{31}	4·20118894
TM_{21}	5·13562230
TE_{41}	5·31755313
TE_{12}	5·33144277
TM_{02}	5·52007811
TM_{31}	6·38016195
TE_{51}	6·41561638
TE_{22}	6·70613319
TM_{12}	7·01558667
TE_{02}	7·01558667

6. CAVITY RESONATOR MEASUREMENTS

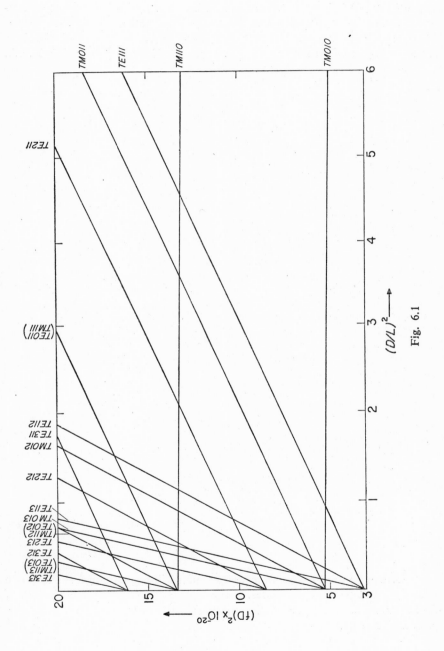

Fig. 6.1

Equation (6·3) is in a convenient form for representing the modes graphically by a series of straight lines with $(fD)^2$ as ordinate and $(D/L)^2$ as abscissa A few lower order modes are shown in Fig. 6.1, and the field configurations of the modes used in the experiments to be described are shown in Fig. 6.2.

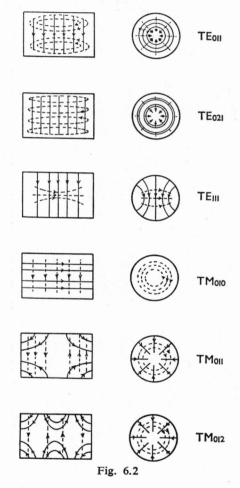

Fig. 6.2

Fig. 6.1 can be used to design a cavity which will resonate in a single mode well removed from other modes. If the frequency is fixed suitable values of D and L can be selected. Equation (6.3) can also be expressed in terms of the wavelength of the radiation by putting $\lambda_{lmn} = c/f_{lmn}$ and it then becomes

$$(1/\lambda_{lmn})^2 = (\Gamma/\pi D)^2 + (n/2L)^2. \tag{6.4}$$

As n is the number of half-waves in the length of the cylinder the right-hand term could be written $(1/\lambda g)^2$ where λg is the wavelength in the cylinder. When the diameter is very large compared with the wavelength the first term on the right-hand side can be neglected and the guide wavelength is equal to the "free-space" wavelength. It is clear for example that the pipe used by Michelson, Pease and Pearson had no effect on the propagation of light. As D becomes smaller a point will be reached when

$$(1/\lambda_{lmn})^2 = (\Gamma/\pi D)^2 = (1/\lambda_{\text{critical}})^2. \tag{6.5}$$

At this value L and the guide wavelength must be infinite, and the wave will not be propagated at all if D is further reduced. This value of λ_{lmn} is sometimes called the critical wavelength, and the corresponding value of f_{lmn} the critical frequency.

3. THE QUALITY FACTORS OF CAVITY RESONATORS

In addition to its resonance frequency an important parameter of a cavity resonator is its quality factor (Q) which determines the sharpness of resonance. This will clearly depend on the conductivity of the walls and on the shape of the resonator. It is calculated from the general expression

$$Q = 2\pi \frac{\text{Maximum energy stored in electric (or magnetic) field}}{\text{Energy lost in one cycle}}.$$

The stored energy is calculated by integrating the quantity $\varepsilon E^2/2$ over the volume of the cavity and the lost energy by using Poynting's theorem and integrating the wall losses over the surface. The solution for cylindrical cavities involves Bessel functions but can be conveniently plotted in terms of $Q\,\delta/\lambda$ against D/L. The results for a few common modes are given in Fig. 6.3.

4. EFFECT OF WALL LOSSES ON FREQUENCY

The main effect of the losses on the resonant frequency is that due to the penetration of the fields into the metal surface thus effectively increasing the dimensions of the cavity. The effect can most conveniently be expressed in terms of the Q factor. Essen and Gordon-Smith (1948) used the analysis given by Bernier (1946) who showed that the propagation constant γ' of a wave-guide of finite conductivity is related to that in a lossless guide by

$$\gamma' = \gamma[1 - \mu\delta A(1-j)/2\sqrt{2}] \tag{6.6}$$

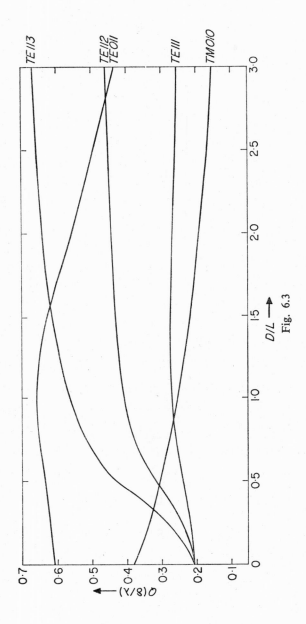

Fig. 6.3

where $\gamma' = 2\pi f'/v$, $\gamma = 2\pi f/v$, A is a function of the field and δ is the skin depth. The imaginary part of this expression corresponds to the attenuation but the real part corresponds to a reduction in frequency. In terms of frequency 6,6 becomes

$$f' = f[1 - \mu\delta A(1-j)/2\sqrt{2}]. \tag{6.7}$$

The measured frequency is the real part of this expression

$$f' = f[1 - \mu\delta A/2\sqrt{2}] \tag{6.8}$$

the quality factor Q can be expressed as the real part of γ' devided by twice the imaginary part of γ' and equation (6.8) therefore becomes

$$f' = f/(1 + 1/2Q). \tag{6.9}$$

This problem has subsequently been treated by Slater (1950). There is also a second order effect because the resonance is slightly distorted but this can be neglected in the experiments described here.

5. ESSEN AND GORDON-SMITH 1947

Essen and Gordon-Smith pointed out that in all previous methods the main error arose not from the measurement of the length or the time interval but from the experimental setting of some parameter such as light intensity to that value which made the distance and time correspond. The setting accuracy was low and in consequence the average of many settings was taken. Under such conditions systematic errors could remain undetected. The close agreement between different results could therefore be misleading. In the case of the cavity resonator method the setting accuracy is high. If the Q of the cavity is 20,000 the current in the cavity falls to 0·5 of its maximum value for a detuning of $f/40,000$. With a reasonably large amplitude of resonance and stable conditions it was therefore relatively easy to obtain a precision of setting of 2 parts in 10^6. The frequency, or time interval, and the dimensions could be measured with an accuracy of 1 part in 10^6. The high precision of setting made it possible to examine and reduce systematic sources of error and there was no longer any useful purpose in averaging a large number of measurements.

5.1. The Cavity Resonator

A number of factors were considered in the choice of the material and dimensions of the resonator. In order to obtain a sharp resonance and a small correction term due to the wall losses a material of high conductivity is required but unfortunately such materials like copper or silver are rather too soft for the best mechanical working and have a high temperature coefficient of expansion. It was decided as the best compromise at the time to construct the first resonator of copper and to use the good workshop facilities available to obtain the highest possible uniformity of dimensions. There was

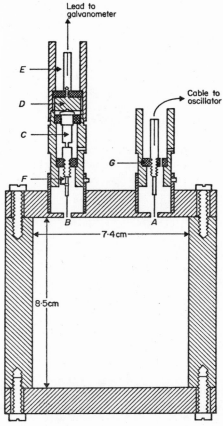

Fig. 6.4

also considerable latitude in the choice of dimensions and of the resonant frequency. Oscillators were available at that time for any frequency up to 10 GHz and as the frequency is increased the penetration of the electric field into the walls decreases. The skin effect correction is therefore smaller. On the other hand oscillators at lower frequencies were more stable and from the point of view of presicion of construction with the honing equipment available the optimum size of resonator was about 8 cm in diameter and length, corresponding to frequencies near 3 GHz. The dimensions were therefore chosen so as to give undisturbed resonances in a number of modes in this frequency range. A cross-section of the cavity is shown in Fig. 6.4. The cylinder was turned from a solid rod and then honed to be uniform, the exact dimensions not being important. The end faces were ground to be flat, square and parallel; the end plates were ground flat and were fixed to the cylinder by eight screws. The coupling probes A and B consisted of wire of 0·015 cm diameter which penetrated less the 0·01 cm through holes of 0·03 cm diameter. It was established experimentally, by using holes of different sizes and different insertions of the probes, that any residual effects were probably less than 0·6 parts in 10^5. The detector loop B led straight to a crystal detector C and to a galvanometer, but in some experiments the crystal detector was removed and a coaxial cable led to a superheterodyne receiver. This method of detection proved to be slightly more sensitive. The sockets E and G were drilled out to permit the free passage of air because the measurements were made with the cavity in a vacuum.

5.2. Procedure of measurement

The arrangement of apparatus is shown in Fig. 6.5. The cavity resonator was mounted in a "bell-jar" which could be evacuated and one probe was connected to the valve oscillator. The types used were at that time known as CV 35 for the E_{010} mode and CV 234 for the E_{011} mode. The types are mentioned because the frequencies of these valves could be varied very smoothly. It was established by experiment that the power from the oscillators did not change significantly when the frequency was changed through the required range. The other probe was connected to a galvanometer if the crystal detector was connected in the probe circuit or to a superheterodyne receiver as shown in Fig. 6.5. The kind used is known as a spectrum

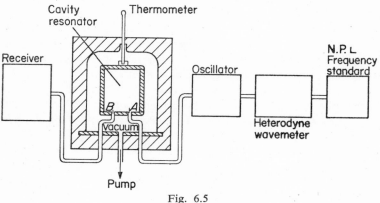

Fig. 6.5

analyser and is useful because the received signal is displayed on a cathode ray tube for a considerable range of frequency without retuning the receiver. One observer set the oscillator to give a maximum deflection on the galvanometer or receiver, and another simultaneously measured the frequency of the oscillator by means of a heterodyne wavemeter designed by Essen and Gordon-Smith (1945). This in turn was related to the N.P.L. frequency standard. The Q of the resonator was measured by observing the change in frequency required to reduce the current to one half of its value. The detector used was a "square law" instrument, and the detected signal was therefore reduced to $1/\sqrt{2}$ of its peak value.

In order to ensure a stable and uniform temperature of the resonator the bell jar was evacuated at least 12 hours before measurements were made, and was heavily lagged. Since the ambient temperature was fairly constant lagging was more satisfactory than a simple form of temperature control.

5.3. Measurement of dimensions

The dimensions were measured in the Metrology Division of the N.P.L. in a temperature controlled room. For the internal diameter a horizontal comparator was used. It had rounded contact tips (0.4 cm radius) and operated with a pressure equivalent to 0.3 kg weight. A suitably wrung combination of slip-gauges and end-pieces served as the reference basis, and differences from this and a diameter were indicated on the optical scale of the comparator which could be read directly to $1.3 \mu m$ and estimated to $0.3 \mu m$. The measurements were made over eight symmetrically disposed diameters at seven positions along the axis of the cylinder. The slip gauges were measured by interferometry in terms of the wavelength standard.

The length was measured by means of a vertical comparator having a rounded contact tip and an optical scale similar to that of the horizontal comparator. One end of the cylinder was placed in contact with the horizontal plate of the instrument and the height of the other end at various positions was compared with a wrung combination of reference standard slip-gauges. Some tests were made to show whether the diameter and length were affected by screwing on the end plates but within the accuracy of the measurements ($\pm 0.8 \mu m$) there was no evidence of deformation. In the routine testing of gauges the standards and the gauges are of similar kinds of steel and temperature effects tend to be self compensating. In this case, however, the cylinder being of copper special precautions were taken and the temperatures of the gauges and the cylinder were continuously recorded. A correction was also made for the fact that the pressure of the gauge produced different depressions on the copper and steel surfaces.

5.4. Results

Preliminary measurements were made using the TM_{010}, TM_{011}, TM_{012}, TE_{011} and TE_{111} modes. As will be clear from Fig. 6.2 the method of coupling used favours the excitation of the TM_{01} modes. The TM_{010} and TM_{011} modes were excited at satisfactory amplitude with the probes flush with the wall of the cavity, but for the others it was necessary to let them protrude into the cavity for 1 or 2 mm and in the case of the TE modes the probes were bent into loops and directed so as to link with the magnetic field. The size of the coupling holes for these experiments was 0.15 cm. The results are given in Table II In view of the penetration of the coupling probes for the three last modes it was decided to use only the first two modes in the final measurements. It was decided also to try and improve the uniformity of both the diameter and length of the resonator. The size of the coupling hole was reduced to 0.03 cm.

TABLE II
Preliminary results

Mode of resonance	Constant Γ	Correction factor $(1 + 1/2Q)$	Measured frequency f' (GHz)	c_0 (km/sec)
TM_{010}	2·404825	1·000028	3·10212	299,797
TM_{011}	2·404825	1·000035	3·56238	299,798
TM_{012}	2·404825	1·000030	4·67881	299,785
TE_{011}	3·831706	1·000015	5·24396	299,799
TE_{111}	1·841184	1·000031	2·95078	299,777

Mean diameter of resonator $7·39759 \pm 0·00012$ cm; variations $\pm 0·00045$ cm; mean length $8·55844 \pm 0·0003$ cm.

TABLE III
Diameter of the cavity resonator expressed as the difference from the mean value of 7·39957 cm. Unit 1×10^{-5} cm.

Axial position from one end (cm)	Diameter at 20 °C in different diametrical planes							
	1	2	3	4	5	6	7	8
0·13	−15	− 6	+ 1	+ 1	+ 7	+ 3	0	−10
1·3	−20	−11	− 4	− 2	+ 1	+ 2	− 4	−16
2·8	−20	−13	−17	− 2	+ 9	+ 7	− 4	−20
4·3	−15	− 9	− 6	+ 3	+ 8	+10	− 3	−14
5·8	− 2	+ 6	+13	+13	+27	+24	+13	0
7·3	− 3	− 3	+ 6	+10	+14	+ 9	+ 7	0
8·4	+ 7	− 1	+ 5	+ 8	+10	+ 8	+ 2	+ 9

The results were based on a series of dimensional measurements made between two series of electrical measurements. The detailed dimensions showing the uniformity achieved are given in Tables III and IV. The cylinder is seen to be slightly oval in shape with a difference of $2\mu m$ (3 parts in 10^5) between the maximum and minimum diameters. Since the field for the modes employed is symmetrical about the axis the effective diameter was taken as the mean value. The length at the inner edge of the cylinder was slightly greater than that at the outer edge and varied smoothly round the circumference by a total of $2 \cdot 2\ \mu m$. The effective length was taken as the mean value at the inner edge.

The measured values of resonant frequency and correction factors and the values of velocity are given in Table V.

TABLE IV

Length of the cavity resonator expressed as a difference from the mean value at the inner edge, 8·53637 cm. Unit 1×10^{-5} cm.

Position of measurement	Length at 20 °C		
	near inner edge	centre	near outer edge
1	− 8	−13	−16
2	+ 5	+ 2	− 8
3	+ 9	+ 6	+ 1
4	+ 9	+ 9	+ 3
5	+ 7	+ 2	− 3
6	− 2	− 5	−12
7	−13	−14	−17
8	− 9	−13	−17

TABLE V

Resonant frequencies of the TM_{010} and TM_{011} modes and values of velocity c_0.

Date	Mode of resonance	Correction factor $(1 + 1/2Q)$	Measured frequency f'_{lmn} GHz	c_0 (km/sec)
2 Oct 1946	TM_{010}	1·000028	3·10125	299,793
2 Oct 1946	TM_{011}	1·000035	3·56380	299,791
25 Oct 1946	TM_{010}	1·000028	3·10128	299,796
25 Oct 1946	TM_{011}	1·000035	3·56377	299,789

Length of resonator $8 \cdot 53637 \pm 0 \cdot 00008$ cm; diameter of resonator $7 \cdot 39957 \pm 0 \cdot 00003$ cm; constant $\Gamma = 2 \cdot 404825$. Mean value of velocity $c_0 = 299,792$

The maximum errors due to various causes are estimated as follows:

1. Setting to resonance and measurement of frequency 4×10^{-6};
2. Uncertainty of temperature of the resonator 2×10^{-6};
3. Errors of dimensional measurements 3×10^{-6};
4. Residual effects of coupling holes and probes 6×10^{-6};
5. Non-uniformity of resonator 10×10^{-6};
6. Uncertainty of Q 5×10^{-6}.

Estimated maximum error = 30×10^{-6}.

The final estimate of the error is obtained by the linear summation of the maximum errors from different causes. A more usual way would be to take the root mean square as is done in determining the standard deviation of a single observation. This would give limits of $\pm 6 \times 10^{-6}$ or ± 2 km/sec. The authors comment that the uncertainties due to the first three causes were established from the precision of repetition, a knowledge of the characteristics of the receiver and the frequency stability of the oscillator, but that the errors due to the other causes could not be estimated with such certainty. Error due to the coupling was the result of an experimental investigation, and that due to non-uniformity from rather general considerations. The maximum deviations of the dimensions from the mean were $+2 \cdot 7$ μm and $—2 \cdot 0$ μm and since they occurred in a gradual manner it was estimated that the average dimensions were known to $0 \cdot 5$ μm. The previous result obtained by Essen (1946) with much less uniform resonators was taken to support the view that the mean value could be used, although it could not be assumed that the non-uniformity of dimensions affected the different modes in the same way. For this reason a rather large error of 1×10^{-5} was allowed. The total value of the correction for the resistance of the walls was approximately 3×10^{-5} and it was stressed that there was some uncertainty because of imperfections in the surface of the resonator and a lack of knowledge of the precise path of the current. Although these were to some extent taken account of by using the measured and not the theoretical value of Q the maximum residual error was estimated to be 5×10^{-6}.

The results obtained before and after the dimensional measurements differed by a larger amount than was usually obtained on dismantling and reassembling the resonator and no explanation could be given for the greater difference between the values obtained for the two different modes. These discrepancies were not pursued further because it was realized that they could best be resolved by using a different type of resonator. The authors state in the introduction to their paper that "the work was done concurrently with an already full programme, and except for the cavity itself existing equipment has been used. As the experiment progressed it became clear that further work with different modes of resonance and with cavities of different types would be of interest".

The enormous improvement obtained by using modern radio frequency techniques is evidenced by the fact that an experiment carried out quickly almost as a side-line to the main programme of work should give a precision of measurement at least 50 times as great as that given by years of painstaking effort with the classical optical method.

6. ESSEN 1950

This experiment was designed to eliminate or at least reduce the two effects which were believed to contribute most to a residual systematic error. In the description of the work considerable emphasis is therefore placed on experimental detail. The two major effects were those of the coupling probes or loops and the non-uniformity of the dimensions. The effect of the coupling was reduced by altering the length of a resonator see (Fig. 6.6) to give a number of successive resonances for which n in equation (6.3) has the values $n = 1, 2, 3, \ldots$. The wavelength in the cylinder was found by taking the differences in the length. The length for the first resonance ($n = 1$) which might be affected by the coupling loops and also by the gap between the plunger and the walls of the cylinder was not included in the calculations. The modes used all belonged to the TE_{01n} series for which the electrical field is circular and does not cross from the end walls to the cylindrical walls. A small gap therefore has very little effect on the resonance. The lengths were measured by means of slip gauges between a ball at the lower end of the piston and a lapped base-plate, the gauges themselves being measured by optical interferometry in a separate experiment. The diameter was measured as before but could also be obtained from measurements of length and frequency. For example if λg is determined at two frequencies using the same mode of resonance (same value of Γ) D can be calculated from equation (6.3)

$$D = \frac{\lambda g_1 \lambda g_2 \Gamma}{\pi} \left[\frac{f_2^2 - f_1^2}{(\lambda g_1 f_1)^2 - (\lambda g_2 f_2)^2} \right]^{\frac{1}{2}}. \qquad (6.10)$$

Alternatively the diameter can be obtained from length measurements alone by using TE_{01} and TE_{02} modes at the same frequency.

D then becomes from equation (6.3)

$$D = \frac{1}{\pi} \left[\frac{\Gamma_2^2 - \Gamma_1^2}{(1/\lambda g_2)^2 - (1/\lambda g_1)^2} \right]^{\frac{1}{2}}. \qquad (6.11)$$

The possibility of avoiding a direct measurement of diameter is important because length and frequency can both be measured more accurately than the internal diameter of a cylinder. As the work developed another important

6. CAVITY RESONATOR MEASUREMENTS

advantage appeared: it enables the skin effect correction which was found to be far from accurate, to be measured and eliminated. The skin effect is automatically eliminated in the length measurements because these depend only on differences and since the diameter is obtained from length and frequency measurements it is also eliminated from this measurement. However as discussed later there may still be some residual error because of experimental limitations.

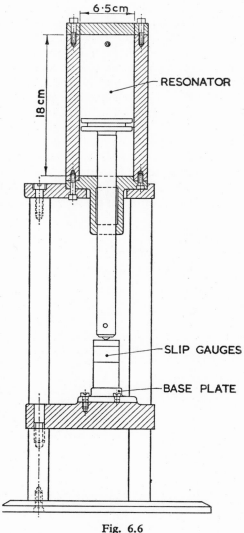

Fig. 6.6

6.1. The cavity resonator

The dimensions of the resonator were chosen carefully so that at frequencies of 9·5 GHz, 9 GHz and 6 GHz a series of resonances having different numbers of half wavelengths could be obtained and each resonance was well removed from the resonances in other modes which are bound to be present in a resonator of this size. For mechanical reasons it was made from case hardened mild steel and the internal surface was ground and honed to be uniform in diameter to $\pm 0.25\,\mu m$. It was then silver plated to a thickness of $5\,\mu m$ and this produced a non-uniformity of $\pm 1.6\,\mu m$ which was greater than expected but still considerably less than that of the copper cylinder used in the previous experiment.

The stem of a micrometer passed through a central hole in the base-plate in order to give a continuous length variation in the initial investigations, but it was not used in any of the final measurements. There was a bar at the lower end of the plunger rod to which wires were attached and passed over pulleys which can be seen in the Fig. 6.7. Counter weights allow the plunger to sink slowly on to the slip gauges and to exert a small constant pressure on them. Four coupling holes of diameter 0·23 cm were spaced evenly round the circumference and small loops were inserted through two of them. The others facilitated pressure equalization when the large cylinder shown in Fig. 6.8 was lowered over the resonator and evacuated.

The internal diameter of the resonator measured by metrological means is given in Table VI.

TABLE VI

Internal diameter of silver plated steel resonator. Values at 20 °C at six symmetrically disposed diametrical planes. Mean diameter 6·51752 cm. Extreme errors $\pm\,0.00003$ cm.

Axial distance from end containing coupling holes (cm)	Difference from mean (Unit 1×10^{-5} cm) Diametrical position					
	1	2	3	4	5	6
0·3	+14	+ 4	− 4	− 4	−19	+ 7
1·6	− 6	− 1	− 4	− 7	− 9	−10
2·8	+ 4	+ 6	− 2	−10	− 1	+ 6
4·1	+ 4	+ 5	− 2	− 8	− 3	+ 4
5·4	+ 4	+ 6	− 4	− 6	− 2	− 6
6·7	+ 7	+ 4	− 8	− 5	− 9	+ 7
7·9	− 1	− 5	−10	− 3	− 9	0
9·2	+ 2	− 3	− 5	− 4	−12	0
10·5	+ 4	−11	−12	− 5	− 6	− 9
12·1	+ 6	+ 3	− 8	− 6	0	− 3
13·0	+10	+ 2	+10	+10	+ 8	− 4
14·0	+ 7	+ 1	+13	+14	+11	+ 2
15·6	+ 1	+ 5	− 2	+ 5	0	− 6

6. CAVITY RESONATOR MEASUREMENTS

The gauges employed in setting the length were measured separately in terms of the length standard by an interferometric method and a check was made to ensure that the movement of the plunger corresponded to that given by the gauges. If the gauges and the plunger were not exactly in line

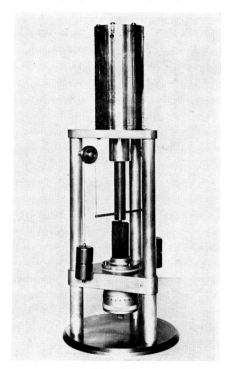

Fig. 6.7

there would be an error given by the secant of the angle between them. The top of the cylinder was removed and the movement of the top of the plunger corresponding to a change in the gauges was measured by direct interferometry (Barrell and Puttock 1950). It was indeed found that the movement was greater than the length of the gauges by $2 \cdot 5 \times 10^{-6} \pm 1 \times 10^{-6}$.

6.2. Measurement of resonant frequency and Q-factors

A microwave frequency standard which generated any frequency in the microwave range accurate to 1 part in 10^8 was available, and the appropriate parts of it were employed in this work. It was used either as the source itself or to measure the frequency of a klystron oscillator which gave a stronger signal in the receiver. Both source and receiver were coupled to the resonator through small coupling loops orientated to give optimum coupling to the TE_{01} modes. The coupling was so small that it did not lower the Q of the resonator or produce any measurable effect on the resonant frequency. Somewhat different techniques were used at the frequencies of 6 GHz and 10·8 GHz to suit the equipment available.

Fig. 6.8

In all cases the resonance was determined from the maximum signal received as the frequency was changed and a power monitor was included to check that the source power did not vary with the frequency change.

The Q-factors were measured by observing the frequency change required to reduce the signal to the half-power points. The values varied from about 30,000 to 120,000 as the length of the resonator increased and were all about 75% of the theoretical value. The precision of setting the frequency was found to be ± 3 parts in 10^7.

6.3. Procedure of measurement

The aim was to obtain a series of resonances such as TE_{011}, TE_{012} ... TE_{01n} at a fixed frequency and so derive the wavelength in the resonator at that frequency, but this was impracticable if fixed gauges were to be used. The nearest convenient gauge lengths required were therefore found by calculation and preliminary experiments, and the exact resonant condition was established by a slight frequency adjustment. The experiment thus yielded n different frequencies corresponding to the n gauge lengths. Each measurement was probably at a slightly different temperature and the values of Q were also different for each mode.

The measurements were made in a temperature controlled room but the cavity resonator itself was not controlled. Thermocouples were fixed at three positions on the cylinder and its supports to ensure that the temperature was uniform and to obtain its value. When the gauges had been wrung on to the base plate and the plunger lowered on to them the cover was lowered and the chamber evacuated. Several hours were allowed to elapse for steady conditions to be reached before any frequency measurements were made. The results were adjusted as follows in order to reduce them to the required form.

1. The frequencies were corrected for the temperature deviation from 20° C. The coefficient found experimentally was $-1 \times 10^{-5}/1°$ C rise and was the same for different positions of the plunger. If the whole apparatus including the gauges had been made from the same material the coefficient would be that of the linear coefficient of the metal, with the sign reversed, but a direct measurement was made because different steels were used for different parts.
2. The correction $f/2Q$ was applied.
3. The lengths, already corrected for gauge errors and the secant error mentioned above, were adjusted to correspond to one frequency near the average of the measured values. The law for this adjustment is, from equation (6.3)

$$\frac{\partial L}{\partial f} = \frac{nf}{2c^2}\left[\left(\frac{f}{c}\right)^2 - \left(\frac{\Gamma}{\pi D}\right)^2\right]^{-3/2} \text{cm/Hz} \qquad (6.12)$$

$1/2Q$ being negligible for this purpose. The numerical value was not required very accurately since the amount of the adjustment was only a few parts in 10^5.

Measurements were made at frequencies near 9·5 GHz and 9 GHz, in order to employ different sets of gauges and an appreciably different wavelength in the resonator. For both of these values the frequency is determined more by the length of the resonator than by the diameter. Measurements were then made at 5·96 GHz, which was the lowest value for which two resonances could be obtained. The frequency in this case was largely determined by the diameter. Finally measurements were made at 10·83 GHz, first using the TE_{01} mode for which the length is more important, and then using the TE_{02} mode for which the diameter is more important.

The resonator was then dismantled for the metrological measurements of its diameter. It was found during this process that the end-faces were slightly concave and they were therefore lapped. The resonator was then reassembled and another set of frequency measurements were taken. The results were based on these two sets of measurements although many preliminary measurements were made in air, and when these were reduced to vacuum values they were all within ± 1.5 km/sec of the final result.

6.4. Results

A typical set of results is given in Table VII, in which the gauge lengths have been converted to cm. In practice the gauges used were in inches and fractions of an inch, the smallest fraction being 0·001 in. The frequencies used before and after the metrological measurements were deliberately made slightly different in order to give different settings of the equipment.

Similar sets of results are given by the author for frequencies of 9·5 GHz at which resonances at eight successive positions were obtained and at 9 GHz at which seven resonances were obtained. The results at 10·83 GHz are not given in detail because owing to the fact that suitable sources and receivers were not available a rather less direct and less satisfactory method of measurements had to be employed and the accuracy was reduced.

The precision of repetition was in this case ± 1 km/sec and although the averages of a number of measurements are included in the summary of results given in Table VIII they are not used in the final result.

In the column 3 of Table VIII a figure is given showing the relative importance of the diameter and length in the result, and in column 4 is the result obtained by using the metrological value of diameter. The discrepancies

TABLE VII

Results for H_{01n} mode at 5·960 GHz and 5·970 GHz

Date	Mode	Corrected gauge length (cm)	Fequency corrected to 20 °C and for Q factor (GHz)	Length adjusted for 5·959 GHz	$\lambda g/2$ (cm)	$c_\bullet$ (km/sec)
8/6/49	H_{011}	9·184621	5·959115	9·182910	7·460465	299,784·5
8/6/49	H_{012}	1·722099	5·958983	1·722445		
		Resonator dismantled, faces trued and reassembled				
				Length adjusted for 5·9698 GHz		
15/10/49	H_{011}	9·301444	5·969900	9·300384	7·344232	299,784·6
15/10/49	H_{012}	1·955720	5·969780	1·956152		

6. CAVITY RESONATOR MEASUREMENTS

are much greater than the precision of measurement and as both frequency and length can be measured with an error less than 1 part in 10^6 it appears that the error occurs in either the value of diameter or of the correction factor $(1 + 1/2Q)$ both of which depend on the surface of the resonator. By using the measurements at 10.83 GHz it would be possible to omit the correction factor and obtain a direct measurement of the effective diameter from equation (6.11). This effective diameter would include the normal surface penetration, any additional penetration due to a layer of insulating or low conductivity material and also any effects due to mechanical imperfections and errors of metrological measurement. Unfortunately the measurements at this frequency proved to be less accurate than the others and are therefore used only to give supporting evidence of the nature of the discrepancy. The final results were calculated in two ways. The correction term $(1 + 1/2Q)$ was included to allow for normal surface penetration and the effective diameter was calculated from equation (6.10), using the results at 9·5 GHz and 5·96 GHz and also at 9 GHz and 5·96 GHz. The values found in this way were 6·51774 cm and 6·51772 cm, respectively, and the average value thus differs from the metrological value by 0·00021 cm. Values of c_0 using this value of diameter are given in column 5. They will be correct if the discrepancy in the diameter is independent of frequency, which would be true if, for example, it is due to a layer of dielectric material or simply to an error of measurement.

On the other hand, if the discrepancy is due to a penetration which varies with frequency then it would be better to correct for it by adjusting the correction term $(1 + 1/2Q)$ to a value which brings the values at different frequencies into agreement. This is found to be $(1 + 2·8/2Q)$ and results based on this are given in column 6.

TABLE VIII

Summary of results

1 Approximate frequency (GHz)	2 Resonant modes	3 $\dfrac{(\Gamma/\Pi D)^2}{(1/2L)^2}$	4 c_0 Using measured diameter (km/sec)	5 c_0 Using calculated diameter 6·51773 cm (km/sec)	6 c_0 Using correction term $1 + 2·8/2Q$ (km/sec)
9·500	$H_{011} - H_{018}$	0·54	299,790·0	299,793·3	299,792·3
9·000	$H_{011} - H_{017}$	0·64	299,789·3	299,792·9	299,792·9
5·960	$H_{011} - H_{012}$	7·8	299,784·6	299,793·1	299,792·6
10·830	$H_{011} - H_{019}$	0·37	299,789	299,792	299,791
	$H_{021} - H_{023}$	8·9	299,785	299,794	299,792

Mean value (ignoring results at 10,830) 299,792·5 km/sec. Maximum error ± 3 km/sec; error as standard deviation ± 1 km/sec.

An inspection of the surface revealed a faintly yellow discolouration together with greyish patches. It is generally believed that the surface tarnish on silver is mainly silver sulphide which is a very poor conductor. The film is detected as a yellowish interference colour when it is about 0·03 μm thick but the action is progressive and the grey colour suggested a considerably thicker layer. It is also commonly known that worked surfaces of silver have a lower conductivity than the bulk metal. There is published evidence of films up to 0·2 μm thick having a resistivity ten times as great as that of the bulk metal, and a film of this nature would increase the surface reactance by more than the value given by the $1/2Q$ term.

It appeared that there were some effects which were dependent on frequency and some that were not and the author decided to take the mean of column 5 and 6 for the final result.

The errors are estimated as follows:

Maximum error in measurement of frequency and length including temperature effects	$\pm 4 \times 10^{-6}$
Maximum error in measurement of diameter by either the metrological or electrical method $\pm 5 \times 10^{-6}$ or as a contribution to error in c	$\pm 2 \times 10^{-6}$
Maximum residual errors due to tarnish estimated as half the difference between columns 5 and 6	$\pm 2 \times 10^{-6}$
Total maximum error by linear addition	$\pm 8 \times 10^{-6}$

If the more usual way of estimating the errors as a standard deviation were adopted the value would be $\pm 3 \times 10^{-6}$ or ± 1 km/sec.

7. HANSEN AND BOL 1950

The only other cavity resonator method completed is that due to Hansen and Bol at Stanford University. No full account of this work appears to have been published and the details included here are taken from a short letter by Bol (1950), review articles by Beardon and Watts (1951) and Mulligan (1952), and a visit made by one of the writers (L. E.) in October 1948. This visit was arranged with Professor Hansen but unfortunately at the time of the visit he was already in hospital with an illness which proved to be fatal. The apparatus was described by Bol and Barclay. The TE_{012} and TE_{021} modes were excited in the cavity resonator sketched in Fig. 6.9. The resonator was made from steel and silver plated on the inside. The diameter was uniform to within ± 3 μm. The actual value of diameter was not measured directly but was calculated from the length and the frequencies of the two modes from equation (6.11). The two resonant frequencies were only a few MHz apart in 3 GHz and presumably the

6. CAVITY RESONATOR MEASUREMENTS

values of λg in equation (6.11) were adjusted to correspond to the same frequency. The cylinder rested on an optically flat base-plate, and the top end face was supported at a small distance above the cylinder on three spherically ended steel rods passing through holes in the walls to the base-plate. The length of the resonator was determined by that of the rods which were measured in a separate experiment by interferometry. The coupling holes were about 3 mm in diameter and the coupling loops were drawn well back in the holes. The effects on the resonant frequency of the coupling holes, the top gap and the non-uniformity of diameter had been calculated

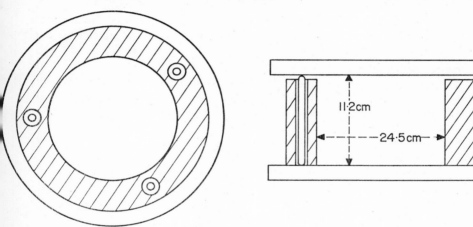

Fig. 6.9

by Hansen and the theoretical values checked experimentally over a range of values. At the time of the visit no result was available.

Bol (1950) gave the result as 299,789·3 ± 0·8 km/sec. He considered the possible error due to the presence of silver oxides or sulphides and allowed 0·5 km/sec for this effect. Bearden and Watts (1951) after consultation with Bol give further details of possible errors as follows:

Contour variations	3×10^{-7}
Error in Q	1×10^{-7}
Frequency measurement	3×10^{-7}
Coupling loop	2×10^{-7}
Spacing between end-plates	5×10^{-7}.

Although the diameter in this experiment was obtained from length and frequency measurements as in Essen's experiment the skin effect is not thereby eliminated, since a single length is used. It is possible that the correction due to this should have been larger than the theoretical value which was applied and that the value given for c_0 was too low in consequence.

8. CAVITY RESONATOR MEASUREMENTS PROPOSED BY ZACHARIAS AND HARRISON

Zacharias and Harrison (1956) proposed an elaboration of Essen's method. Although in this method the effects of imperfections and resistance of the walls are largely eliminated, experimentally there could be a residual effect if the tarnishing of the surface for example was not uniform. In the experiment proposed a number of frequencies were to be used so that the nodes and antinodes of the standing wave would occur at a large number of places along the cylinder and thus average out the effect of such non-uniformities. A number of technical improvements were made. The cylinder is of fused quartz, very accurately ground and coated with silver, and the movement of the end plate is measured directly in terms of the wavelength of the mercury 198 green line by optical interferometry. The accuracy aimed at is 1 part in 10^8 and it is therefore also necessary to express the frequency in terms of the atomic unit. Progress in this work has been reported but at the present time (1968) no result has been published.

CHAPTER 7

Radar, Spectroscopic and Quartz Modulator Measurements

1. ASLAKSON 1949

Radar techniques were developed for detecting enemy aeroplanes and for directing bombers over their targets. The position is obtained from the travel time of a radio pulse and the application of the method to distance measurement was naturally attractive to the geodesist.

Aslakson (1951) reports that a test programme by the U.S. Army Air Forces was concluded in 1947. A total of 47 distances between 100 km and 600 km in the Caribbean area were measured and the results were analysed. One of the conclusions reached was that they gave evidence of a possible error in the accepted value of the velocity of propagation of radio waves. It was felt at first (Aslakson 1949a) that it would be presumptuous to question the accepted value for the velocity of light, but in view of the results obtained by Essen and by Bergstrand whose work is described in Chapter 9 the measurements were later used as the basis for a new determination of c.

The method is basically very simple. The time of travel of a pulse of radio waves to a distant aeroplane and back again is recorded directly on a time scale derived from a frequency standard. The frequency of the standard can be chosen so as to make the time scale read directly in terms of distance. Timing pulses and the radar pulses produce vertical deflections of the cathode ray tube trace. The radar pulses and the horizontal movement of the time base are repeated at regular intervals in synchronism, so that the time markers remain stationary while the radar pulse returned from a moving target moves steadily along the scale as the distance changes. In order to obtain sufficient resolution only a small part of the time or distance scale is displayed, for the determination of the position of the radar pulse between two markers. The whole number of time markers elapsing between the transmission and reception of the radar pulse is counted electronically.

There must be a line of sight path between the transmitter and aeroplane and a high radio frequency transmission should be used to minimize the effect of the ground on the propagation velocity. The range is therefore limited by the curvature of the earth and the height of the plane. If the height is 12 km a radar system such as Shoran used by Aslakson can measure a distance of 400 km, and if the plane flies midway between two ground stations the range can be doubled (see Fig. 7.1a).

The Shoran system operates on frequencies near 300 MHz. Pulses of waves, about 1 μ sec in duration are sent from the plane to two stations A and B where they are received and retransmitted to the plane. The delays in the instruments are measured separately and allowed for, the actual propagation time being recorded as a distance on the receiving equipment in the aeroplane. As the plane crosses the line between A and B the individual distances and also the sum of the distances vary approximately as a parabola (Fig. 7.1b), the minimum of which corresponds to the cross-over point. A flat minimum is not easy to observe and in practice the plane described a figure of eight (see Fig. 7.1c). The individual distances to A and B then varied linearly during the cross-over, and a plot of the sum of the distances enabled the minimum of the parabola obtained to be determined with a good precision. The figure of eight was then repeated in the reverse direction in order to reduce instrumental errors and systematic errors due to the observer's reaction time. There were many other errors to be considered. The wave travels through a path of varying refractive index and in consequence the velocity is reduced by a varying amount and the ray is bent. In the work described by Aslakson (1949b) an additional aeroplane flew at different heights and positions of the path to measure the pressure, temperature and humidity of the air.

It was also found that there was a large error in Shoran measurements due to the fact that the slope of the received pulse varied with signal intensity. An empirical relationship was found for the error due to signal intensity and this was applied to the results. The spread of the results was thereby reduced and this was accepted as evidence of the validity of the correction. The average error corresponded to a change in velocity of about 15 km/sec and if it had not been applied a result close to the old optical value would have been obtained.

For the purpose of a velocity determination six of the lengths which had also been measured by geodetic triangulation were selected, and the result obtained was

$$c_0 = 299{,}792 \cdot 4 \pm 2 \cdot 4 \text{ km/sec} .$$

The total spread of the six measurements was 10·5 km/sec but it should be remembered in considering the precision that each measurement was the result of about 240 individual distance readings, and that the individual values are obtained by a least squares triangulation adjustment as in geodetic triangulation.

2. FLORIDA SURVEY 1951

The U.S. Air Force completed another survey in Florida in 1950 using an improved Shoran equipment. In particular the troublesome signal intensity error was effectively removed. Shoran measurements were made

7. RADAR, SPECTROSCOPIC AND QUARTZ MODULATOR MEASUREMENTS

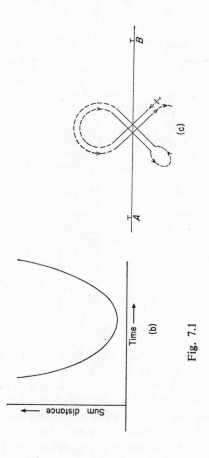

Fig. 7.1

TABLE I

Comparison of Shoran with geodetic distances

Line 1	2 Geodetic	Distances in miles 3 Shoran based on $c_0 = 299{,}776$ km/sec	4 Shoran based on $c_0 = 299{,}794\cdot2$ km/sec
2—6	40·6131	40·6041	40·6066
3—4	96·7171	96·7049	96·7108
5—6	100·3098	100·2988	100·3049
5—1	118·9953	118·9840	118·9912
1—6	133·0113	132·9985	133·0066
2—3	134·9698	134·9546	134·9628
5—2	139·1225	139·1127	139·1211
3—6	145·8427	145·8276	145·8365
1—2	145·8884	145·8768	145·8857
4—6	190·5047	190·4889	190·5004
2—4	199·1914	199·1735	199·1855
5—3	226·9903	226·9698	226·9835
5—4	235·5264	235·5042	235·5185
1—3	277·0569	277·0347	277·0515
1—4	320·1519	320·1271	320·1466

of 15 lines and were compared with the values found from geodetic triangulation. The results are given in Table I.

It is seen from columns 2 and 3 that to a large extent the discrepancy is proportional to the distance and would therefore be removed by the adoption of a different value for c_0. From a least squares analysis Aslakson determined the best value for c_0 and also the best value for a systematic error not depending on distance. In column 4 the distances are calculated using the new value of c_0. The value obtained was

$$c_0 = 299{,}794\cdot2 \pm 1\cdot4 \text{ km/sec}$$

and Aslakson believed it to be much more reliable than his previous value.

3. OTHER RADAR MEASUREMENTS

In the United Kingdom a number of measurements were made to check the velocity of radio waves in radar applications. Smith, Frank in and Whiting (1947) used propagation frequencies between 22·9 MHz and 59·5 MHz using two paths of lengths 125 km, mainly over sea, and 140 km, over hilly country. One of the ground stations was used for the transmission of the pulses which were received and retransmitted on a different carrier frequency from that used by the other ground station. The accuracy claimed was only ±50 km/sec and it was concluded that there was no significant difference over the two paths, and that when allowance was made for an average value of refractive index there was no significant difference from a value for c_0 of 299,776 km/sec. Jones (1947) used the Oboe blind bombing system operating on a frequency near 3 GHz. As before, the transmitter and transponder were both on the ground but there was an optical path between them. The distances involved were about 60 km. It was concluded from the experiments that the velocity of radio waves at ground level was 299,687 ± 25 km/sec. The refractive index of air at radio frequencies is so dependent on the moisture in the air that there is little point in relating this value with c_0 without information concerning the atmospheric conditions.

Jones and Cornford (1949) also used the Oboe system but with the transmitter in an aeroplane as in Aslakson's experiments. The plane flew across a line which was in this case an extension of the line joining two ground stations, and when it was directly above the line the travel time of the radio pulses was a minimum. The experiment was carried out with the plane flying at three different heights. It was concluded that the most probable values for the mean velocity between the ground and an aircraft at 3000 m, 6000 m and 9000 m are 299,713 km/sec, 299,733 km/sec and 299,750 km/sec respectively. The experimental results if extrapolated to ground level are about 11 km/sec higher than the calculated value, but no comments were made on the cause of the discrepancy.

An extensive radar survey was carried out in Canada during 1949—1953 (Ross, 1951, 1954) using Shoran equipment which was calibrated assuming a value for c_0 of 299,776 km/sec and a refractive index at ground level and standard conditions of 1·0002885. It was found that the value of c_0 determined by Essen, Bergstrand and Aslakson gave more consistent results and the value of refractive index was also increased to that found by Essen and Froome. The details given by Ross show the extent of the corrections for delays in the instrument, refractive index slope and plane height. The results reported for the line between Boundary and Churchill are given in Table II.

It is seen that the corrections amount to 2 parts in 10^3 so that for an accuracy of 2 parts in 10^5 they must be determined to 1%.

The results obtained by Aslakson gave support to the higher values which were being obtained but the errors inherently associated with the radar method prevent it from being considered today as a useful method for measuring c_0. The long distances involved cannot be measured in terms of the length standard with the same precision as a base line or distances in a laboratory, the corrections for instrument delays cannot easily be determined with the accuracy required and the refractive index correction for a path of varying height is difficult to compute. The radar method is to be regarded more as a valuable application of the velocity of light to geodetic purposes.

TABLE II

Corrected altitude of aeroplane	20,290	20,290 feet
Antenna altitude	1,310 feet	109 fee
Slope distances	147·7150 miles	149·1050 miles
Delay	− 0·1904	− 0·1734
Velocity	+ 0·0083	+ 0·0075
Frequency	− 0·0006	− 0·0006
Indicator	+ 0·0010	+ 0·0010
Shoran distance corrected	147·5333	148·9395
to sea level	− 0·0760	− 0·0725
for slope	− 0·0438	− 0·0490
for curvature	+ 0·0082	+ 0·0082
	147·4217	148·8262
Sum distance		296·2479
Eccentricity of antenna		+ 0·0024
Reduced Shoran length		296·2503 miles

4. SPECTROSCOPIC METHODS

The frequency of a spectral line is given by the Bohr equation

$$f = (E_1 - E_2)/h \qquad (7.1)$$

where E_1 and E_2 are the energies of the atomic or molecular states and h is Planck's constant. If the wave emitted by an atom as a result of the transition is propagated in a space of refractive index n the wavelength of the radiation is

$$\lambda = c_0/fn . \qquad (7.2)$$

The value of c_0 could therefore be obtained simply by multiplying the frequency and vacuum wavelength of the radiation. Such measurements could be carried out on a radio frequency spectral line but there would be no advantage in doing this. Any stable radio frequency oscillator can be used equally well because its frequency in terms of the standard can be measured more accurately than the wavelength. In the optical region the wavelengths of spectral lines can be measured but not their frequencies. It is however possible to derive c_0 by measuring the spectroscopic constants of a suitable line in terms of wavelength and frequency.

The rotational energy levels of a diatomic or linear polyatomic molecule are given to a good approximation by

$$E_J/h = F(J) = BJ(J+1) - DJ^2(J+1)^2 \qquad (7.3)$$

where J is the rotational quantum number and B and D are rotational constants. Transitions for which the values of J differ by ± 1 give the pure rotational spectrum of the molecule. The frequencies of the lines can be measured in the microwave region and have the values

$$f = F(J+1) - F(J) = 2B(J+1) - 4D(J+1)^3. \tag{7.4}$$

For the $J = 0 \to J = 1$ absorption transition this reduces to

$$f = 2B - 4D. \tag{7.5}$$

The rotational lines can be observed in the infra-red region as transitions between two vibration levels. The constants may be slightly different for the higher and lower states but those for the lower state are assumed to be the same as those in the microwave region. The wavelengths are given by

$$1/\lambda = fn/c_0 = 1/\lambda_0 + (B' + B)m + (B' - B - D' + D)m^2 - \\ - 2(D + D')m^3 - (D - D')m^4 \tag{7.6}$$

where the quantities are expressed in cm^{-1} units. λ_0 is the wavelength of the central band, and $m = J + 1$ for the R branch of the band and $-J$ for the P branch. The values of B and D can be found from a least squares adjustment of the results obtained for a number of lines. D is only a small correction term of the order of $10^{-6} B$ and an error in the value of c_0 used to convert it to frequency units is not important. This value of D can then be used in equation (7.5) and c_0 obtained from the two values of B

$$c_0 = B \text{ microwave}/B \text{ infra-red}. \tag{7.7}$$

Corrections must of course be made for the refractive index n under the conditions of measurement.

The basic limitation of this method is clear from equation (7.6). The value of B is obtained from two values of $1/\lambda$ which differ on the average by less than 0·5%. If the accuracy of the length measurements is 5 parts in 10^7 the accuracy of measuring B is thus reduced to about 1 part in 10^4 although some systematic errors may cancel out in the least squares evaluation.

This method was first used by Rank, Ruth and Vander Sluis (1952) who examined the rotation bands of HCN, and associated their infra-red value of B with the microwave value measured by Nethercot, Klein and Townes (1952). The value obtained was

$$c_0 = 299{,}776 \pm 6 \text{ km/sec}.$$

In their first measurements they used a grating and although it was one of the best that could be obtained, wavelengths could not be measured to better than 5 parts in 10^7. Rank, Shearer and Wiggins (1954) therefore repeated the experiments using a purely interferometric method and obtained the value

$$c_0 = 299{,}789{\cdot}8 \pm 3 \text{ km/sec}.$$

In addition to the error arising from the fact that the value of the constant depended on the difference between two nearly equal wavelengths, the error in the value of the centre of the band was not negligible. In 1955 Rank, Bennett and Bennett (1956) made a wavelength determination of the same

TABLE III

Values of $1/\lambda$ of the rotational lines of CO at $4.7\,\mu m$

		Unit 1/cm				
	J	Measured	Measured − calculated	J	Measured	Measured − calculated
R	23	2224·694	0·004	14	2086·322	0·000
	22	2221·732	0·004	15	2082·009	0·007
	21	2218·733	0·005	16	2077·650	0·001
	20	2215·685	−0·004	17		
	19	2212·600	−0·012	18	2068·851	0·006
	18	2209·498	0·001	19		
	17	2206·345	0·002	20	2059·911	−0·001
	16	2203·147	−0·006	21	2055·391	−0·006
	15	2199·929	0·005	22		
	14	2196·661	0·003	23	2046·271	−0·001
	13	2193·357	0·002	24	2041·663	0·002
	12	2190·010	−0·004	25	2037·030	0·010
	11	2186·636	−0·001	26	2032·349	0·002
	10	2183·226	0·003	27	2027·635	−0·007
	9	2179·761	−0·011	28	2022·899	−0·009
	8	2176·287	0·003	29		
	7	2172·759	−0·001	30		
	6			31		
	5	2165·602	−0·001	32	2003·659	−0·001
	4			33	1998·767	−0·005
	3	2158·309	0·006	34	1993·867	0·013
	2	2154·596	−0·003	35	1988·910	0·004
	1			36	1983·919	−0·010
	0			37	1978·923	0·002
P	1	2139·432	0·002	38	1973·871	−0·013
	2	2135·554	0·004	39	1968·805	−0·013
	3	2131·639	0·004	40		
	4			41		
	5	2123·770	−0·002	42	1953·459	0·014
	6	2119·677	−0·007	43	1948.252	−0·010
	7			44		
	8	2111·555	0.009	45	1937·831	0·019
	9	2107·413	−0·012	46	1932·555	0·011
	10	2103·265	−0·017	47	1927·246	−0·002
	11	2099·096	0·012	48	1921·926	0·001
	12	2094·870	0·007	49		
	13	2090·603	−0·006	50		
				51	1905·776	−0·010

band of HCN using the mercury 198 green line as the reference standard. This yielded a value for c_0 of

$$c_0 = 299{,}791 \cdot 9 \pm 2 \text{ km/sec}.$$

The same method was employed by Plyler, Blaine and Connor (1955) with the infra-red spectrum of CO. Although a grating was used for the spectrometer, a wider range of wavelengths was available so that the overall accuracy of the determination of the constants was probably improved. Forty-three lines in the absorption band at $4 \cdot 67$ μm were measured together with fifteen lines having values of J between 32 and 52 in the emission spectrum. The values of $1/\lambda$ are given in Table III together with the difference from values calculated from the least squares adjustment. The polynomial

$$1/\lambda = 1/\lambda_0 + (B_1 + B_0)\, m + (B_1 - B_0)\, m^2 - 4Dm^3 \qquad (7\ 8)$$

was fitted to each set of results and the values of $1/\lambda$, B_1, B_0 and D_0 were determined. The results obtained from two sets of measurements are given in Table IV together with the microwave value of B_0 obtained by Gilliam, Johnson and Gordy (1950) and Bedard, Gallagher and Johnson (1953). These spectroscopic measurements lack the precision of the more direct

TABLE IV

Constants of CO

$1/\lambda_0$	$2143 \cdot 272 \pm 0 \cdot 003$	$2143 \cdot 275 \pm 0 \cdot 002$
B_1	$1 \cdot 90501 \pm 0 \cdot 00007$	$1 \cdot 90498 \pm 0 \cdot 00004$
B_0	$1 \cdot 922528 \pm 0 \cdot 00007$	$1 \cdot 922521 \pm 0 \cdot 00004$
B_0 (microwave)	$57 \cdot 63565$ GHz	
D_0	$6 \cdot 26 \pm 0 \cdot 042 \times 10^{-6}$	$6 \cdot 25 \pm 0 \cdot 03 \times 10^{-6}$
	$c_0 = 299{,}792 \pm 6$ km/sec	

methods involving simply the measurement of frequency and length but they provide interesting evidence that the microwave and infra-red values of B are in fact related exactly by c_0.

5. QUARTZ MODULATOR METHOD

If light is passed through a piece of quartz crystal along the optic or Z axis and an electric force is applied along the electric or X axis the plane of polarization of the emergent beam will be rotated. If the applied electric field E varies at the frequency f and the light is passed through a quartz crystal plate placed between two Nicol prisms the resultant intensity is

$$I = I_0 \sin^2 (klE \sin 2\pi ft) \qquad (7.9)$$

where l is the path length in the crystal and k is a constant of the crystal. The crystal thus acts in a similar manner to the Kerr cell. If f is made equal to a natural vibrational frequency of the quartz plate the amplitude of its oscillations increases enormously and so does the electro-optical effect. There is now an additional effect due to the fact that the mechanical strains in the crystal change its refractive index. If it is excited in an overtone mode there are bands of different refractive index and the plate behaves like a diffraction grating. This effect was used by Houstoun (1941, 1944) for measuring the velocity of light in water and he later (1950) applied it to measure-

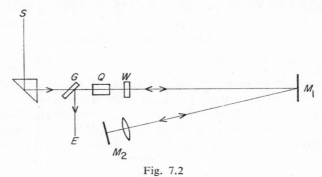

Fig. 7.2

ments in air. The arrangement he used is shown in Fig. 7.2. A parallel beam of light was passed through the glass plate G and the quartz plate Q to a mirror M_1, 19 m away, from which it was reflected to the lens and mirror M_2. It was reflected back along the same path to the quartz plate and eye piece E. The dimensions of the quartz plate were $3.51 \times 11.03 \times 12.975$ mm and it vibrated at its 135th overtone mode at a frequency of about 106 MHz. The first order spectrum was deviated by about 35 min of arc. The equipment was set up using the direct image. The end of the quartz was bevelled so as to deviate the direct image by 35 min, and the deviation was removed by the glass wedge W. When the oscillator was switched on and the wedge W removed the first order spectrum was obtained. The mirror M_2 was moved to give a minimum intensity which occurred when the path distance was $(n + 1/2)\lambda/2$. The value of n was 55 and was thus not in doubt. The setting to a minimum which was about one third of the intensity of the maximum was not easy to obtain and in the words of the author "to say that its determination gives no feeling of aesthetic satisfaction is an understatement". The final result was obtained from 20 measurements each one being the mean of 20 settings. The spread of the measurements was still 180 km/sec. The result was $299,698 \pm 9$ km/sec giving, when corrected for refractive index,

$$c_0 = 299,775 \pm 9 \text{ km/sec}.$$

7. RADAR, SPECTROSCOPIC AND QUARTZ MODULATOR MEASUREMENTS

McKinley (1950) described some measurements made in 1937 using a quartz plate in its fundamental mode. When the quartz was used as modulator and detector he obtained a probable error of the mean of 100 observations of the order of 2%. A big improvement was obtained by using the quartz crystal as modulator and a photocell as detector, and he obtained a result

$$c_0 = 299{,}780 \pm 70 \text{ km/sec.}$$

CHAPTER 8

Microwave and Radio-wave Interferometer Methods

1. INTRODUCTION

The merits of the optical interferometer for the precision measurement of length or, conversely, the radiation wavelength, are well-known and require no elaboration here. But an interferometer operating with a radiation whose frequency can be measured offers an extremely accurate method for the determination of the velocity of this radiation which is then derived from the product of the measured wavelength and frequency.

It is possible to measure the frequency of electromagnetic waves produced by a suitably stabilized oscillator well into the microwave region comprising centimetric, millimetric and sub-millimetric wavelengths, but even today it is not possible to measure the frequencies of optical lasers: thus the wavelength of the radiation used in microwave determinations of the velocity of "light" have been much larger than that of optical radiation.

Fundamentally there is no difference between an interferometer for light waves and one for microwaves. In both, some arrangement of movable reflector varies the phase of the radiation returned by it to a detector, relative to the phase arriving by some other route from the same source. The cavity resonator described in Chapter 6 is similar to a Fabry-Perot optical interferometer and if the diameter of the cavity is made very large the analogy becomes more exact. In the Michelson interferometer radiation reflected from a fixed mirror placed in one arm interferes with that reflected from the movable mirror in the other arm. The detector can be the eye, a photographic emulsion, or a photo-electric cell. Displacement of the movable mirror causes the detector output to fluctuate, and in optical interferometry this fluctuation repeats exactly for every half-wave movement of the mirror. In the microwave region of frequencies the detector may be a crystal and galvanometer or a superheterodyne receiver; but diffraction causes an error in apparent wavelength measurements deduced from the mirror displacements, so that some special method for the elimination of this effect is required.

It is not possible, in practice, to use an interferometer operating at microwave frequencies in exactly the same manner as its optical counterpart. The ratio of lens or mirror aperture to wavelength is responsible for the difference. For optical interferometry one would not think of working with components of dimensions under, say, 10^4 wavelengths. With the microwave oscillators at present available components of dimensions comprising this number of wave-lengths would be very large. Consequently it is inevitable that the wavefront of any microwave radiation emitted or reflected by a component of practicable size must, owing to diffraction, change its shape progressively with distance from the component.

Nevertheless the microwave interferometer is one of the most powerful methods at present available for the determination of the velocity of electromagnetic radiation. Although the apparent wavelength as measured by such an instrument is a function of the dimensions of its components, it is possible to assess relatively accurately the limits of error introduced by the correction for this effect, since the total deviation of the measured wavelength from the true free-space wavelength can be made quite small.

2. FROOME'S DETERMINATION OF 1952

Figure 8.1 shows in diagrammatic form all the essential details of the interferometer (Froome 1952). The source of microwaves was a stabilized klystron oscillator of frequency 24·005 GHz, corresponding to a wavelength of about 1·25 cm. The output from the valve entered the hybrid junction ('magic-T') A where it was divided into two equal parts. One part was

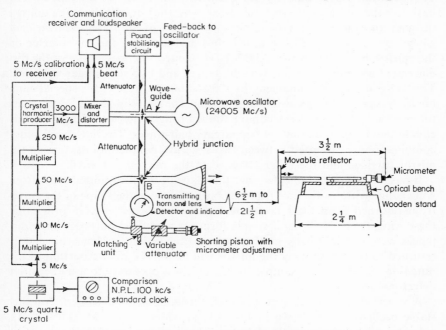

Fig. 8.1. Micro-wave interferometer for velocity determination.

required to operate the stabilizing circuit; the other entered the second hybrid junction B, constituting the beam divider of the interferometer.

At B the energy was once more split into two equal parts. Half passed through the transmitting horn out into the open, and was ultimately returned to the horn after reflexion from the distant movable reflector. This represened the "variable" arm of the interferometer.

The "fixed" arm was a wave-guide equivalent of the variable arm. In this, the other half of the energy from B passed along a wave-guide through a "matching" unit and then a variable attenuator, finally to be reflected by an adjustable shorting piston. The function of the matching unit was to balance the reflections in the other arm occurring at the front surface of the transmitting horn, the unit being placed at the same electrical distance from the junction B as the horn mouth. Consequently the matching was insensitive to frequency. Thus both arms of the interferometer reflected microwave energy back to the hybrid junction B, half being passed to the detector and half towards the oscillator. Interference between the beams reflected in the open and fixed arms took place in the wave-guide containing the detecting device. The intensity of the beam from the fixed arm could be balanced against that from the open arm by adjusting the variable attenuator, so that if the phase in the open arm was altered by moving the reflector the detector recorded zero wherever the phases of the two beams were opposite. This occurred (approximately) for every half-wave shift of the mirror. The basis of the wavelength measurement was thus to displace the mirror through an exact number of minima by means of accurately known end-gauges plus a very small (up to 0·1 mm) movement of the micrometer on the mirror bench shown in Fig. 8.1. To control the magnitude of the displacement a combination of two end-gauges and two slip-gauges was used. The end-gauges, of nominal lengths 100 and 60 cm, were of the special combination type used in precision engineering. The two slip-gauges, one wrung to each end of the 160 cm combination, brought the total length closely equal to an integral number of "apparent" half-waves. The flat-ended gauge combination was inserted between two spherical contact points, one on the mirror carriage and the other attached to the micrometer anvil.

It has already been indicated that the shape of the transmitted wave-front will change with its distance from the horn. Likewise the shape of the wave-front returned from the reflector will change with distance. In order that these wave-fronts shall be of simple form for analytical purposes large distances were required between the horn and the reflector. Hence the energy returned to the detector by the mirror will be only a minute portion of the outgoing radiation, for example 10^{-4}, and it was necessary to use a sensitive detector.

A superheterodyne receiver was used throughout for detection. Another stable oscillator (not shown in Fig. 8.1) provided energy at 24·050 GHz, a crystal mixer placed in the detector "arm" of the junction B converting the difference between this frequency and the signal (24·005 GHz) into an intermediate frequency of 45 MHz. This was amplified, and the rectified signal was indicated by two micro-ammeters in series. One of these meters was placed near the transmitting horn, the other on the mirror bench adjacent to the micrometer. With this device and with a distance of about 21·5 m between the horn and the mirror, the precision of a single setting to a minimum was about ± 3 μm, even with the smallest reflector of dimensions 15×15 cm. (These were also the dimensions of the aperture of the transmitting horn used throughout the experiment.)

In all measurements the mirror was displaced through 259 minima, corresponding to approximately 162 cm. The mirror displacements were made at five different distances between the mirror and the transmitting horn to obtain sufficient data for the elimination of diffraction effects. These distances were 6·5, 9, 12, 16 and 21·5 m, and refer to the position of the movable mirror when the combination gauge was withdrawn.

Three different square aluminium reflectors were used, with sides of 15, 23 and 30 cm. The 15 cm mirror was used at all distances, the 23 cm mirror at all except 6·5 m, and the 30 cm mirror at the three greater distances.

The temperature of the gauge combination was ascertained from three thermometers reading to 0·01° C suitably placed with their bulbs in contact with the surface of the gauge. The length of the gauge at 20° C was determined by a method based on optical interferometry.

The measurements made in air were reduced to the vacuum condition by means of the refractive index formulae given in equation (3.10). The air temperature was ascertained to an estimated accuracy of $\pm 0.05°$ C and the barometric pressure to ± 0.1 mm. The water-vapour pressure was determined to ± 0.1 mm by means of two electrically operated Assman hygrometers.

Each wavelength measurement for a given transmitting horn to mirror separation had to be made in such a manner as to eliminate the small disturbing effect of stray reflections. These can be grouped into four classes. First, there are reflections back into the horn from fixed objects well removed from the mirror bench; these could be ignored because they are not altered by the action of making a mirror displacement. Secondly, there was scattered radiation reaching the horn from objects in the neighbourhood of the mirror; this was serious because it affected the shape of the reflected wave-front differently at the limits of the mirror displacement. It was eliminated by making a number of mirror displacements each one based on a datum position (or initial mirror bench micrometer setting) about 0·08 cm in advance of its predecessor. In this way the phase of the disturbance was varied in relation to the phase of the wanted reflection, and the measured wavelength underwent a cyclical variation. Thirdly, there was scattering from the mirror to some fixed object and then back into the horn, and *vice versa*. The experiment was actually conducted in a room so large (27 m long $\times$ 21 m wide $\times$ 20 m high) that only the floor might be troublesome in this connexion. Reflections from the floor were minimized by arranging scaffold planks on edge, so tilted and inclined that they absorbed or scattered in a random manner away from the horn-mirror axis. This axis was also set skew to the walls of the room so that the effect of the end wall (behind the mirror bench) which gives rise to reflexions of the first and second class was as small as possible. The perturbations arising from floor reflections will also tend to be eliminated by the technique of making measurements at a number of horn-mirror distances.

The fourth error was that due to multiple reflections between the horn and the mirror, the horn reflecting an appreciable fraction of the energy returned to it from the mirror. Thus multiple reflections could produce a small error in the displacement measurements. Their effect was eliminated by the same technique used for the removal of the second class of reflection error so that the phase of the modulation of the principal reflected beam, produced by the multiple reflections, was systematically varied through 2π. In the worst case the cyclical variation found from the two combined sources of error was ± 1 in 10^5.

The technique of frequency measurement was relatively straightforward, not differing greatly in principle from that used by Essen in the cavity resonator measurements. As indicated in Fig. 8.1, the basis was a 5 MHz quartz crystal oscillator which could be compared directly with an N.P.L. frequency standard. The 5 MHz output was scaled up to exactly 24 GHz by direct frequency multiplication to 250 MHz followed by two stages of harmonic generation. The final stage also mixed with a small fraction of the output of the microwave oscillator at 24·005 GHz, to produce a "beat" at 5 MHz which could be meassured by a communications receiver. The frequency of the klystron oscillator could be ascertained and held constant by this method to 1 part in 10^8.

The interferometer operated in the Fraunhofer diffraction region because the mathematical analysis of the near-zone or Fresnel diffraction region is very difficult owing to the rapid changes of wave-front shape that occur in this case.

In the Fraunhofer region the distance between transmitting horn and movable reflector is so large that both may be regarded almost as point sources of radiation. If they were true point sources the wave-front from the horn reaching the mirror would be a sphere of radius z, equal to their perpendicular distance apart. Since the horn and mirror dimensions were several wavelengths in extent the energy re-radiated after reflection from the mirror had an additional retardation (over and above z) dependent upon the difference in shape between the spherical wave-front and the plane face of the mirror. This difference was a function of the depth of a segment of a sphere of radius z intercepted by the sides of the mirror aperture. Similarly, the reflected wave upon reaching the horn also had a spherical

8. MICROWAVE AND RADIO-WAVE INTERFEROMETER METHODS

front of radius z. The phase of the radiation conveyed to the wave-guide was dependent upon the summation of this wave-front over the mouth of the horn, thus giving rise to another additional retardation dependent upon the spherical segment intercepted by the horn aperture.

Let the total retardation of the wave returned to the horn, expressed as a length, be z_r; then the foregoing physical picture may be given mathematical expression in the following form

$$z_r = 2z + \frac{f(h) + f(m)}{z}. \tag{8.1}$$

$f(h)$ represents some unknown function of the horn dimensions and $f(m)$ a similar function of the mirror dimensions. $\{f(h) + f(m)\}/z$ will hereafter be referred to as the asymptotic form of the diffraction correction.

The method used for evaluating the experimental results consisted in solving equation (8.1) by a least-squares method, after modification by the addition of a relatively complicated term. This modification made allowance for departures from the asymptotic form which arose from deviations of the wave-fronts from true spheres. The determination of the additional term involved integration of a complex variable over six dimensions and thus entails a considerable amount of computational work. It is opportune to mention here that the average diffraction correction applied to all the wavelength measurements was equivalent to -12 km/sec, of which -0.5 km/sec was due to the deviation from the asymptotic form of the correction.

To evaluate the correction term the total retardation z_r was calculated for the various mirror-horn distances used and for variously assumed dimensions for the horn and the mirror apertures. (These aperture dimensions were not necessarily the same as the true geometrical dimensions.) This involved calculating the field distributions at the mirror, the reflected field caused by these at the horn mouth, and finally the integral of this reflected field over the horn mouth. The phase of the reflected wave entering the throat of the horn depends upon this integral.

For the full mathematical treatment the reader is referred to the original paper, only the results will be quoted here. It was shown that the interference equation is

$$N\lambda/2 = (z_1 - z_2) - K\left(\frac{f(\delta_2)}{z_2} - \frac{f(\delta_1)}{z_1}\right). \tag{8.2}$$

Here N is the number (259) of interference minima in a mirror displacement $(z_1 - z_2)$ of about 162 cm, K is a constant and $f(\delta)$ is the calculated deviation from the asymptotic form of diffraction correction. For a given mirror the measured displacements for the different horn-mirror distances

TABLE I

Diffraction corrections applied to observer B's measurements

	Horn-mirror distance (gauges removed)	1st assumption: apparent horn and mirror apertures equal to geometrical dimensions			2nd assumption: horn aperture 10×10 cm; mirror apertures two wave-lengths larger than 1st assumption in height and breadth		
		Least-squares diffraction correction μm	Calculated correction μm	Velocity (km/sec)	Least-squares μm	Calculated μm	Velocity (km/sec)
15 cm mirror	6·5 m	160·8	182·5	mean for 15 cm mirror 299,793·2	163·8	162·6	mean for 15 cm mirror 299,793·0
	9·0	78·1	88·4		79·9	79·3	
	12·0	41·9	47·4		43·2	42·8	
	16·0	22·9	26·3		23·4	23·3	
	21·5	12·6	14·4		12·9	12·8	
23 cm mirror	9·0	134·6	136·3	mean for 23 cm mirror 299,792·2	135·3	134·8	mean for 23 cm mirror 299,792·2
	12·0	74·9	75·1		74·3	74·0	
	16·0	41·1	41·7		41·7	41·5	
	21·5	23·3	23·0		23·0	22·9	
30 cm mirror	12·0	116·8	109·8	mean for 30 cm mirror 299,791·9	114·5	114·3	mean for 30 cm mirror 299,792·3
	16·0	67·5	62·7		64·5	64·3	
	21·5	37·8	34·6		35·6	35·5	

8. MICROWAVE AND RADIO-WAVE INTERFEROMETER METHODS

are substituted in equation (8.2) and a solution obtained for λ by the method of least squares. $f(\delta)$ is given values dependent upon the assumptions made regarding the effective horn and mirror sizes. The diffraction corrections for the different measurements obtained in this way were compared with the entirely calculated correction. If the least-squares and calculated corrections did not agree, fresh values were assumed for the horn and mirror dimensions to derive new values for $f(\delta)$; these new values of $f(\delta)$ are inserted in equation (8.2) and the least-squares solution reworked. This method of successive approximation is not so arduous as might appear at first sight. Since $f(\delta)$ differed little from unity, the process converged rapidly, as will be seen by reference to Table I, showing the measurements of one observer (B) and the related calculations.

From these results it is seen that no further assumptions after the second could possibly affect the resulting velocity. Indeed, whatever were the exact distributions over horn and mirror apertures, it is apparent that no appreciable change in velocity would result from their usage. The agreement between the least-squares experimental corrections and those calculated from theory for the second assumption above is quite remarkable. The other two observers showed in general somewhat greater scatter in their results.

This basic process for the evaluation of the diffraction correction is not essentially different from that given in detail for the analysis of the 72 GHz four-horn interferometer in section 3.

The least-squares method of solution of equation (8.2) does not, of course, eliminate systematic errors in the mirror displacements $(z_1 - z_2)$. The fact that each mirror tends to give an answer slightly different from the others showed that the eight mirror displacements comprising a single wavelength measurement were not quite sufficient to eliminate completely the effects of the various unwanted reflections mentioned in the previous section.

The estimated error of ± 0.05 km/sec in the diffraction correction (see Table IV) has been derived by assessing the probable effect of the various approximations made in the derivation of $f(\delta)$ when put into the least-squares solution of equation (8.2).

Table II gives, for each observer, the measured displacements $(z_1 - z_2)$ reduced to vacuum conditions.

The displacements quoted in Table II lead to the values of velocity given in Table III.

The arithmetical mean of all the values of c_0 in Table III is 299,792·62 km/sec, for which the standard deviation is ± 0.68 km/sec. When the latter is combined statistically with the possible systematic errors listed in Table IV the standard deviation is raised to ± 0.72 km/sec. The final value of the free-space velocity in vacuo was

$$c_0 = 299{,}792{\cdot}6 \pm 0{\cdot}7 \text{ km/sec},$$

as previously recorded (Froome 1952).

TABLE II

Mirror displacements (unit, 1 cm) reduced to vacuum

	Mirror-horn distance* (cm)	Observer A	Observer B	Observer C
15 cm mirror	648	161·745 45	161·745 94	161·745 75
	900	36 80	37 77	37 77
	1204	33 95	33 75	34 29
	1601	31 77	31 90	32 00
	2146	30 96	30 98	31 09
23 cm mirror	900	42 37	42 56	41 80
	1204	36 59	36 84	36 59
	1601	33 36	33 33	33 37
	2146	31 46	31 29	31 27
30 cm mirror	1204	40 84	40 51	40 56
	1601	35 82	36 15	36 12
	2146	32 49	32 49	32 38

* Gauges withdrawn

TABLE III

Values of c_0 (unit, 1 km/sec) derived from Table II

	Mirror-horn distance (cm)	Observer A	Observer B	Observer C
15 cm mirror	648	299,793·27	299,792·91	299,793·40
	900	2·14	3·32	3·73
	1204	3·38	2·67	3·90
	1601	2·84	2·91	3·21
	2146	3·21	3·16	3·41
23 cm mirror	900	2·34	1·93	2·31
	1204	2·64	2·64	3·16
	1601	2·47	2·17	2·80
	2146	2·32	1·86	2·14
30 cm mirror	1204	1·49	1·99	1·65
	1601	1·95	3·17	2·88
	2146	1·39	1·75	1·41

The length of the combination gauge used was obtained by an optical interference method, using wavelength standards of a mercury-198 isctope lamp. The length determined by mechanical comparison with standards derived directly from the British National copy of the metre agreed with the optical determination to 2 parts in 10^7.

TABLE IV

Estimated subsidiary uncertainties

Nature of error	Corresponding uncertainty in c_0 (km/sec)
Pressure of water vapour in atmosphere (± 0.1 mm)	± 0.18 (6 parts in 10^7)
Length of gauges + micrometer error	± 0.09 (3 parts in 10^7)
Temperature of gauges ($\pm 0.02°$ C)	± 0.06 (2 parts in 10^7)
Uncertainty in the coefficient of expansion of the gauges	± 0.05 (1.5 parts in 10^7)
Diffraction correction	± 0.05 (1.5 parts in 10^7)
Refractive index formula: dry air	± 0.03 (1 part in 10^7)
water vapour	± 0.03 (1 part in 10^7)
Air temperature ($\pm 0.05°$ C)	± 0.02 (5 parts in 10^8)
Barometric pressure (± 0.1 mm)	± 0.01 (4 parts in 10^8)

3. FROOME'S DETERMINATION OF 1958

The principal merit of the moving reflector type of interferometer just described is its basic simplicity of operation. In other respects the design is very unsymmetrical and also wastes available radiation. Furthermore, the use of a mirror complicates the evaluation of the diffraction correction. The four-horn interferometer removes those disadvantages and has one further benefit; this is that a uniform expansion of the room and equipment does not affect the position of an interference minimum. In practice this produced a four-fold gain in thermal stability over the interferometer described in the previous section. The four-horn instrument was also a two-beam interferometer working in the Fraunhofer diffraction region.

Two versions of the symmetrical interferometer have been investigated. The first (Froome 1954) was an experimental prototype operating at 24 GHz, and used for a thorough study of the diffraction correction; the second (Froome 1958) was a very high precision version operating at 72·006 GHz corresponding to a radiation wavelength of 4 mm and this instrument only will be described here, the theory being given in full.

As before, the basis of the experiment consisted of the simultaneous measurement of the free-space wavelength and frequency generated by the microwave source. An additional innovation was the use of a cavity resonator refractometer for the direct determination of the refractive index of the air in the neighbourhood of the interferometer. This refractometer (Froome 1955) operating at approximately 72 GHz was in principal the same as those used by Essen and Froome (1951) at 24 GHz in their first measurements of the refractive index of air and its principal constituents.

The interferometer (Figure 8.2) employed a movable carriage supporting a pair of receiving apertures (horns of 8 × 6 cm rectangular aperture fitted with polystyrene lenses) at opposite extremities, symmetrically placed between an identical pair of transmitting horns energized with microwave

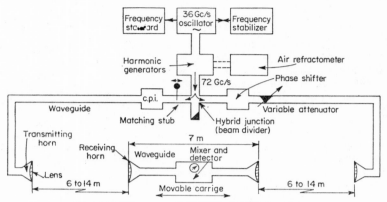

Fig. 8.2. Diagram of 72 GHz interferometer.

power conveyed by waveguide from a common source. The axes of the four horns were carefully aligned and the motion of the carriage was also arranged parallel to this direction. The two received signals were mixed together to produce interference, the resultant output undergoing a minimum for every half-wave displacement of the carriage. The separation between each transmitter and the corresponding receiving horn was of the order of 10 m and as the effective aperture dimensions were squares with sides of about 6 cm, the receiving horns were well within the Fraunhofer diffraction field of the transmitters. The received wavefronts were thus approximately spherical and of radius equal to the appropriate transmitting-receiving horn separation. It was the change of these radii with the position of the movable receiving carriage that gave rise to the diffraction correction, the evaluation of which was the primary function of the 24 GHz prototype.

The movable carriage of the new interferometer was displaced through 970 half-waves, (a distance of about 2 m) by means of end-standards known in terms of a lightwave standard of length. The value of the microwave wavelength so obtained, when multiplied by the air refractive index and the microwave frequency, gave a vacuum phase velocity which had still to be corrected for the effect of diffraction before the true free-space value could be derived. The magnitude of this correction was found by making seven velocity measurements, each for a different transmitting-receiving horn separation and fitting the results to the theoretical form.

By virtue of (a) the greater carriage displacement, (b) the shorter wavelength and (c) the improved location, all major error-producing influences arising from the method of microwave transmission, propagation and reception were reduced relative to their effect on the prototype by more than one order of magnitude. It is interesting to note that the greatest single uncertainty in the whole measurement arises from the use of the length-standards.

Figure 8.2 shows in diagrammatic form all the essential details of the interferometer and associated equipment. The source of microwaves was the first harmonic of a Pound stabilized Q-band klystron oscillator operating at 36·003 Gz. The greater part of the output (about 10 mW) from this oscillator was fed by means of a waveguide switch into one of two silicon crystal distorter units tuned for maximum harmonic output at 72·006 GHz. One harmonic generator was used to supply the interferometer itself, the other for operating the refractometer by means of which the refractive index of the air in the neighbourhood of the equipment could be measured. A smaller portion of the klystron power was used for driving the Pound frequency stabilizer and a minute fraction for frequency measurement.

The measurement of microwave frequency was accomplished by comparing the klystron output with a high harmonic of the 5 MHz quartz crystal standard. The 5 MHz was multiplied in stages of two to five times up to 600 MHz and then fed into a silicon crystal harmonic generator mounted in a waveguide, so that the harmonic at exactly 36 GHz could be mixed with the klystron output. The beat frequency between the two was detected by means of a calibrated communications receiver. When the refractometer was being used the klystron frequency had to be variable between 36·002 and 36·014 GHz (instead of constant at 36·003 GHz) and was set to the nearest 100 kHz calibration point on the communications receiver corresponding to approximate resonance of the refractometer cavity, the setting to perfect resonance then being made by means of a small tuning plunger attached to the cavity. The accuracy of frequency measurement was good to at least 1 part in 10^8.

When the interferometer was being used to make a wavelength measurement the 72·006 GHz output from the appropriate harmonic generator was guided to a hybrid junction which served as beam divider, and eventually through two long arms of cylindrical waveguide to the transmitting horns. The matching-stub to the left of the beam divider, together with the constant phase waveguide interferometer (c.p.i.), constituted a device for altering the amplitude of the energy transmitted down this arm without producing a phase displacement (Froome 1954). The phaseshifter to the right, together with the variable attenuator, was required in order to adjust the position and balance of the first interference minimum on the movable receiving carriage.

The central platform of the receiving carriage held three wheels arranged so they ran on a cast-iron bed of 3 m length taken from a precision measuring machine. Beneath the surface of the bed ran a counterweight equal in mass to the carriage, but arranged to move in the opposite direction so as to maintain a constant moment on the concrete block supporting the bed and on the floor of the room beneath the block (Fig. 8.3).

The whole apparatus was installed in a very large wind-tunnel room. Across the approximate centre of the room was a "honeycomb" anti-turbulence wall and this was used to form a convenient datum point for measure-

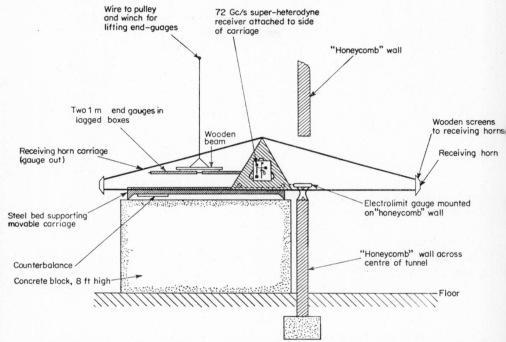

Fig. 8.3. Method of mounting the receiving horn carriage of the 72 GHz microwave interferometer in the duplex wind tunnel room.

ments of displacement of the receiving carriage. A part of the carriage actually passed through a hole cut in this wall. Figure 8.4 is a photograph of the receiving carriage showing the "honeycomb" wall.

A highly sensitive electrical measuring head "Electrolimit" which is divided in units of 0·25 μm was mounted upon the "honeycomb" wall in such a manner as to be entirely independent of the steel bed (and concrete block) supporting the carriage, except when a small ball mounted on the centre of the carriage (in line with the receiving horn axes) was in contact with the sensing anvil of the Electrolimit head. Provision was made for an independent slow motion adjustment of the carriage.

The energy picked up by the receiving horns was transmitted along an oversize rectangular waveguide to a hybrid junction mixer mounted on the central platform of the carriage. The "mixed", or interference signal, was detected by means of a superheterodyne receiver involving a second stabilized Q-band klystron and a crystal harmonic generator as "local oscillator". This was fixed to the side of the movable carriage. The intermediate frequency, chosen to be 60 MHz, was first amplified by a wide-band preamplifier and then by a communications receiver with a final band-

Fig. 8.4. A photograph of the receiving carriage showing the "honeycomb" wall.

width of 5 kHz. The output from the detector stage of this receiver was indicated by a microammeter placed close to the receiving carriage slow-motion adjustment, and the minimum precision of setting was about 0·5 μm. When the interferometer was not in use, the output from the second Q-band klystron could be conveyed by means of a waveguide switch to the detector associated with the refractometer.

The 2 m length-standard used for measuring the carriage displacement through 970 half-waves consisted of two 1 m end-bars set up in line with slip-gauges wrung to each face and separated in the centre by a 6 mm steel ball. When the end-standard was "inserted" into position (i.e. the interferometer was set on the 971st minimum) wheels attached at the Airy points of each gauge automatically and accurately located it on rails fixed to the 3 m bed so that it was free to move longitudinally as the setting on the 971st minimum was made. The diameter of the steel ball and the slip gauges were measured directly by optical interferometry. The 1m end-gauges were compared against a special standard the length of which had been determined by optical interferometry. At this time the best-known optical wavelength was the red line of the cadmium spectrum so this was the basic length standard. A useful secondary standard was a mercury 198 isotope lamp.

The auxiliary constant phase waveguide interferometer has been elaborated in detail in the paper by Froome (1954) describing the prototype four-horn interferometer; but for the purpose of clarity a further summary of its mode of operation is desirable. The basic theory is very simple: if two

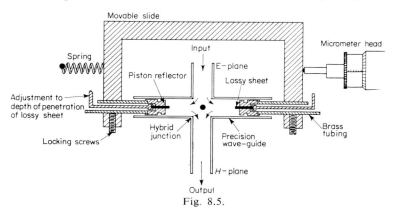

Fig. 8.5.

equal vectors, initially parallel (equal waves in phase), are rotated in opposite directions through the same angle, the direction of their resultant is unchanged while its amplitude is reduced.

Figure 8.5 is a diagram of the c.p.i. It consists of a waveguide interferometer utilizing a hybrid junction for the dual purpose of beam division and recombination. Energy from the beam divider of the four-horn interferometer enters the E-plane arm of the c.p.i. and at the centre of the junction divides into two equal parts and enters the arms containing the mechanically linked movable-piston reflectors. The waves reflected from these pistons interfere at the centre of the junction and normally some energy is transmitted through the H-plane output arm and the remainder is returned through the input arm. In Fig. 8.2 the matching stub between c.p.i. and the main interferometer beam divider prevents further reflexion of this returned energy from the main beam divider. Thus, it is apparent that when the c.p.i. pistons are so adjusted as to give the maximum transmitted output, a rotation of the c.p.i. micrometer in either direction will cause a reduction of output energy. When the movable carriage of the principal interferometer was in the "gauges inserted" position (i.e. set on the 971st minimum) and the amplitude of the waves picked up by the receiving horns was balanced by a clockwise rotation of the c.p.i., the c.p.i. micrometer was said to be at "position A". Anticlockwise rotation to produce amplitude balance led to "position B". If the c.p.i. micrometer was moved from one position of maximum transmission to the next the phase of its output was changed by π so that, for equal output, two new positions (A^1 and B^1) of the micrometer were now possible when the device was used for less than maximum transmission. The c.p.i. was constructed from optical flats of stainless steel completely uniform over the movement of the pistons. When used in the manner described no error arising from its use could be detected.

In order to make wavelength measurements free from all disturbing influences, the following experimental procedure was adopted:

1. The observer, after having performed a preliminary run to adjust (by means of the slip gauges) the 2 m combination end-standard to be closely equal to 970 half-waves, set the movable receiving carriage in contact with the sensing anvil of the Electrolimit gauge mounted on the "honeycomb" wall.
2. The c.p.i. having been set at a position of maximum transmission (the mean of the A and B positions), the variable attenuator and the phase-shifter (see Fig. 8.2) were adjusted to give zero-detector current at the position of the first minimum.
3. The temperature of the end-bar assembly was taken.
4. The observer made four settings on the first minimum, reading the Electrolimit gauge each time.
5. The receiving carriage was run to the limit of its movement and the end-bar assembly lowered into position between the Electrolimit head and the fiduciary ball on the carriage.
6. The c.p.i. was turned to position A and the observer noted the Electrolimit gauge reading for four settings on the 971st minimum.
7. Operation (6) was repeated for the position B of the c.p.i.; the mean of the Electrolimit readings for the A and B positions being taken as the true position of the 971st minimum.
8. The length-standard was removed and the operation (4) repeated to correct for drift, after which the temperature of the length-bar was again taken.
9. The c.p.i. micrometer was moved to the next position of full transmission (defined as the mean of the A^1 and B^1 positions) and the Electrolimit head moved in its clamp by $\lambda/4$ so that the position of the 1st minimum was again within the range of the Electrolimit gauge.
10. Operations (3) to (8) were repeated using the c.p.i. in the A^1 and B^1 positions instead of the A and B positions.

By taking the mean of the two wavelength measurements so far made, the effects of multiple reflexions between transmitting and receiving horns were eliminated, together with all the c.p.i. errors except one. The operations were continued:

11. The refractive index of the air was determined by shutting off the slow air circulation through the refractometer cavity and measuring the change of frequency upon evacuation.

12. A quarter-wave spacer was inserted in the waveguide of the output arm of the c.p.i. The two receiving horns were rotated through 180° about their (horizontal) geometrical axes, thereby reversing the effect of any differences between microwave and geometrical axes.

13. Operations (1) to (10) were repeated. The mean of all the four wavelength measurements thus made was then free from all errors arising from the microwave system, with the exception of that due to diffraction and scatter from fixed objects including the floor, ceiling and walls of the room.

14. The effect of scatter was removed as follows: the axial position of the transmitting horns was altered in steps amounting only to a few wavelengths in total so that the phase of the direct transmission to the receiving horn was varied in a cyclical manner relative to the scattered radiation. For each step operations (1) to (13) were repeated.

15. To determine and eliminate the effect of diffraction, fully corrected (i.e. operations 1 to 14) wavelength measurements were made for seven different transmitting-receiving horn separations.

As before, the diffraction correction arises from the fact that the wave retardation at any point on a receiving horn is slightly in excess of the geometrical separation (z). The total retardation is the integral over the equiphase surface of a receiving horn and approaches the segment of a sphere for the "asymptotic" case as z becomes large. Were the wave-forms truly spherical by the time they reached the receiving horns, the free space wavelength in air (λ) would be related to the carriage displacement by the interference equation

$$N\lambda/2 = \Delta z + A(1/z_1 - 1/z_2). \tag{8.3}$$

N is the number of interference minima in the carriage displacement Δz and A is a constant dependent upon the sizes of transmitter and receiver. The interferometer was always used with the transmitter-receiver distance (z) equal on both sides when the carriage was midway between the terminal minima. Thus z_1 is the larger value of z (and z_2 the smaller) when the carriage is at either limit of its displacement ($\Delta z = z_1 - z_2$).

If equation (8.3) is divided by $N/2$ and multiplied by the product of frequency f and refractive index n, one obtains the equation

$$c_0 = c_0' + B(1/z_1 - 1/z_2), \tag{8.4}$$

where $B = (2Anf)/N$ is another constant, c_0 is the true free-space vacuum phase velocity and c_0' is the measured or "apparent" vacuum velocity.

8. MICROWAVE AND RADIO-WAVE INTERFEROMETER METHODS

Ideally, therefore, we can find both c_0 and B experimentally from equation (8.4) by a simple least-squares procedure involving the measurement of c'_0 for a number of different values of z_1 and z_2. As the diffraction correction must always be small compared with c, z_1 and z_2 need only be measured to the nearest half-wave-length.

In practice, equation (8.4) is unsatisfactory for an accurate evaluation of c_0 because the values of z required are too large for the sensitivity demanded of the apparatus. Accordingly, we modify the spherical or asymptotic formulation of equations (8.3) and (8.4) to permit the use of rather smaller values of z. We put

$$N\lambda/2 = \Delta z + A(\delta_1/z_1 - \delta_2/z_2), \qquad (8\cdot5)$$

where δ is a correction for departure from the initial asymptotic form. δ is equal to the ratio of a calculated additional retardation to the asymptotic term of the same calculation. (This calculated retardation is based upon reasonable assumptions concerning the distributions of electric and magnetic fields at the mouth of the horns.) As $z \to \infty$, δ rapidly approaches unity. If, therefore, conditions are chosen so that δ never differs from unity by more than a small percentage (e.g. not more than 5%), and if the experimental results at a number of different z's are fitted to equation (8.5) by the method of least squares, then the derived value of c_0 is relatively insensitive to the nature of the field assumptions. These are regarded as satisfactory when reasonable agreement is obtained between the least-squares values of A, and the purely theoretical value of A derived on the basis of the field assumptions.

It is seen that this procedure is identical with working a least-squares solution of either of the following equations,

$$N\lambda/2 = \Delta z + K(D_{z_1} - D_{z_2}), \qquad (8\cdot6)$$

$$c_0 = c'_0 + 2Knf(D_{z_1} - D_{z_2})/N. \qquad (8\cdot7)$$

Here K is a constant to be determined by the least-squares method; D_z is the additional retardation of the received wave (in addition to z) calculated on the basis of the field assumptions. These assumptions are adjusted until the least squares result gives $K=1$, the best values of c_0 then being obtained. (D_z will, of course, have an inverse variation with z for its asymptotic term.)

In the following treatment the full mathematical analysis is given. It proves that for chosen experimental conditions the diffraction correction can be evaluated from simple scalar theory. It is only appropriate to treat the case when all four horns are identical because then the most favourable general conditions are achieved.

Consider a transmitting horn (rectangular mouth of sides $2a$ parallel to the x-axis and $2b$ parallel to the y-axis) situated at the plane $z = 0$, and facing a receiving horn of the same aperture size situated at z. The z-axis passes through the centres of the horns. Thus $(x, y, 0)$ is any point in the plane of the transmitting aperture, and (x, y, z) is any point in the plane of the receiving aperture.

It is wished to derive an expression for the phase retardation (in addition to z) of the received wave in order to obtain c_0 from a least-squares solution of equation (8.7).

First calculate the amplitude and phase of the radiation field from the transmitting horn at the plane of the receiver. The vector formulation of what is essentially Huygens' principle is of satisfactory rigour, and Schelkunoff's (1939) statement of this method particularly lends itself to an accurate assessment of the diffraction correction. The whole radiating surface of the horn (i.e. the aperture and possibly the outer boundaries of the horn mouth) is replaced by an electric current sheet of density J and a "magnetic-current" sheet of density M, where

$$\boldsymbol{J}_T = \boldsymbol{\phi} \times \boldsymbol{H}_T, \tag{8.8}$$

$$\boldsymbol{M}_T = -\boldsymbol{\phi} \times \boldsymbol{E}_T. \tag{8.9}$$

Here $\boldsymbol{E}_T$ and $\boldsymbol{H}_T$ are the field vectors of the transmitted wave just outside the boundary of the horn, and $\boldsymbol{\phi}$ is a unit vector normal to the surface of the sheets.

Schelkunoff shows that

$$\boldsymbol{E}_T = \boldsymbol{E}_O + \boldsymbol{E}_{OR}, \quad \boldsymbol{H}_T = \boldsymbol{H}_O + \boldsymbol{H}_{OR}, \tag{8.10}$$

where $\boldsymbol{E}_O, \boldsymbol{H}_O$ are the fields which would exist in the plane of the aperture where the horn continued to infinity ($\boldsymbol{E}_O$ and $\boldsymbol{H}_O$ are the only known fields), and $\boldsymbol{E}_{OR}, \boldsymbol{H}_{OR}$ are the fields reflected by the actual discontinuity of the mouth. From equations (8.8) and 8.9) the "free-space" magnetic vector potential A and the electric potential F are established, where

$$\boldsymbol{A} = \left(\frac{1}{4\pi}\right) \int\int_S \left(\frac{\boldsymbol{J}_T}{r}\right) \exp\left(-ikr\right) dS, \tag{8.11}$$

$$\boldsymbol{F} = \left(\frac{1}{4\pi}\right) \int\int_{S_T} \left(\frac{\boldsymbol{M}_T}{r}\right) \exp\left(-ikr\right) dS. \tag{8.12}$$

The integration should extend over the whole radiating area (S_T) of the transmitter. $k = 2\pi/\lambda$ and r is the distance between any point on the transmitting surface to any point situated in the receiving surface. The radiated fields at these latter points (in the absence of the receiver) are then given by

$$\boldsymbol{E} = -i\omega\mu\boldsymbol{A} + (1/i\omega\varepsilon)\,\mathrm{grad}\,\mathrm{div}\,\boldsymbol{A} - \mathrm{curl}\,\boldsymbol{F}, \tag{8.13}$$

$$\boldsymbol{H} = -i\omega\varepsilon\boldsymbol{F} + (1/i\omega\mu)\,\mathrm{grad}\,\mathrm{div}\,\boldsymbol{F} + \mathrm{curl}\,\boldsymbol{A}, \tag{8.14}$$

8. MICROWAVE AND RADIO-WAVE INTERFEROMETER METHODS

where $\omega = 2\pi f$, ε = electric permittivity of air and μ = magnetic permeability.

Equations (8.13) and (8.14) are rigorous if E_T and H_T are known, but only E_O and H_O are known in practice. Strictly, therefore, if E and H are to be deduced from E_O and H_O, and integration is to be limited to the horn mouth, the "free-space" form of the vector potentials should be replaced by a formulation involving Green's tensor functions in order to preserve the correct field relationships at the boundary of the horn mouth. Such a procedure was avoided by the adoption of suitable experimental conditions.

Part of the experiments performed with the 24 GHz prototype (Froome 1954) were undertaken to determine whether E_{OR} and H_{OR} were of appreciable magnitude.

Hence first consider the fields given by equations (8.13) and (8.14) when it is assumed that $E_{OR} = H_{OR} = 0$ and that the integration is limited to the geometrical aperture of the horn. For the mode of propagation used in the horns the transverse field vectors are mutually perpendicular and the small longitudinal component of magnetic field can be neglected since it is non-radiative (i.e. its integrated effect over the horn mouth is zero), taking E_O parallel to the y-axis, the radiated electric field is given by

$$E_y = ik' \bar{A} - \frac{\partial \bar{A}}{\partial z} - \frac{k'}{ik^2} \frac{\partial^2 \bar{A}}{\partial z^2} \tan^2 \theta, \qquad (8.15)$$

$$E_x = -\frac{k'}{ik^2} \frac{\partial^2 \bar{A}}{\partial z^2} \tan^2 \theta, \qquad (8.16)$$

$$E_z = \frac{k'}{ik^2} \tan \theta - \frac{\partial \bar{A}}{\partial z} \tan \theta, \qquad (8.17)$$

where $\bar{A} = \frac{1}{4\pi} \int_{S_{TA}} \frac{E_O}{r} \exp(-ikr) \, dS$, and integration extends only over the aperture (S_{TA}) $k' = 2\pi/\lambda'$, where λ' is the wavelength at the actual position of the plane of the horn mouth, if the horn continued indefinitely. Since, from waveguide theory

$$\lambda'/\lambda = [1 - (\lambda/4a)^2]^{-1/2} \qquad (8.18)$$

this ratio differs from unity by only 2×10^{-4} and k' can be put equal to k in the amplitude terms.

θ is an angle to be used solely for the purpose of estimating errors and can thus be loosely defined as effectively one-quarter the angle subtended by the receiving horn at the centre of the transmitter, i.e. for this purpose θ is taken to be the larger of $a/2z$ or $b/2z$. The $\tan^2\theta$ terms arise from the gradient of the divergence of the vector potentials and conditions must always be adjusted to keep them negligible (i.e. $\tan \theta < 0.02$).

Another and more important criterion is that $kz \gg 1$. This implies that the transmitter-receiver separation must always exceed 100 λ and arises from the process of differentiation of the vector potentials. If $kz \gg 1$, this differtentiation is straightforward and gives rise to small additional terms in quadrature with the potentials themselves, reduced in magnitude by the ratio $1/kz$. The experimental conditions were so chosen that for the smallest transmitter-receiver separation the disturbance of the diffraction correction by these quadrature terms was of magnitude similar to the accuracy of setting on an interference minimum. Then H can be obtained from equations (8.15) to (8.17) by putting $H = kE/\omega\mu$.

$$\frac{\partial \bar{A}}{\partial z} = -\left[\left(\frac{ik}{4\pi z}\right)\exp(-ikz)\right]\left[\int_{S_{TA}} \frac{E_o \exp[-ik(r-z)]\,dS}{1+(r-z)/z}\right] \cdot \left[\exp\left(-\frac{ik}{k^2 z}\right) - \frac{d}{z^2}\right], \tag{8.19}$$

where d is the asymptotic term (i.e. the $O(1/z)$ term) of the argument of the integral in (8.19), d/z^2 corresponds to the obliquity factor of the simple Kirchoff formulation of Huygens' principle, to which form the equations are now tending. From (8.15) to (8.17)

$$E_y = \left[\frac{i}{4\pi z}\exp(-ikz))\right]\left[\int_{S_{TA}} \frac{E_o \exp[-ik(r-z)\,dS}{1+(r-z)/z}\right] \cdot \left[k' + k\exp\left(-\frac{ik}{k^2 z}\right) - \frac{kd}{z^2}\right], \tag{8.20}$$

$$E_x = 0, \tag{8.21}$$

$$E_z = \left[\frac{i}{4\pi z}\exp(-ikz)\right]\left[\int_{S_{TA}} \frac{E_o \exp[-ik(r-z)]dS}{1+(r-z)/z}\right][k'+k]\tan\theta. \tag{8.22}$$

Thus the diffraction retardation now differs little from the argument of the integral.

Subject to the conditions already imposed, the amplitude of the wave accepted by the receiving horn is given by

$$E_R = i\left[1 + \frac{k}{k'}\exp\left(-\frac{ik}{k^2 z}\right)\right]\left[\int_{S_{RA}} RE_y\,dS\right]. \tag{8.23}$$

where S_{RA} signifies integration over the aperture of the receiver and R is a receiving coefficient, which, because the receiving horns are identical with the transmitters, can be put equal to E_o for the purpose of evaluation of the phase of E_R. This approximation neglects the effect of E_z (equation 8.22) upon E_R. E_z arises from the fact that away from the common receiver-transmitter axis the Poynting vector of the transmitted field is not quite parallel to this axis but inclined to it by an angle whose averaged size is closely equal to θ. Errors in ignoring E_z are only of the order of $\sin\theta \tan\theta$ and are thus very small. The exponent in equation (8.23) has been included because the quadrature (i/kz) term in equation (8.19) must again enter when a horn is used as receiver.

Thus, in equations (8.6) and (8.7)

$$D_z = -(1/k)(\arg E_R + kz). \tag{8.24}$$

So far only horns which have "plane-wave" lenses fitted have been considered, that is E_0 is real. Take two aperture distributions of E_O:

Case 1. $E_O = E_{ax}$ = constant. This corresponds to uniform illumination of the horn apertures, when used as transmitters. E_{ax} is the axial value of E_O. Then

$$D_z = [(a^2 + b^2)/3 + 0.0253 \, \lambda^2 (2k)/(k + k')]/z - [0.289(a^4 + b^4) + \\ + 0.111 \, a^2 b^2 + 0.6127(a^6 + b^6)/\lambda^2]/z^3 + 0(1/z^5 + \ldots) \tag{8.25}$$

Case 2. $E_O = E_{ax} \cos(\pi x/2a)$. This is the theoretical form of aperture distribution for the TE_{01} mode of propagation used. Then

$$D_z = [0.18943 \, a^2 + b^2/3 + 0.0253 \, \lambda^2 (2k)/(k + k')]/z - [0.104 \, a^2 + 0.289 \, b^4 \\ + 0.063 \, a^2 b^2 + (0.1795 \, a^6 + 0.6127 \, b^6)/\lambda^2]/z^3 + 0(1/z^5) + \ldots \tag{8.26}$$

Now consider the effect of phase distortions at the apertures. E_O is no longer entirely real, and the expansion for D_z contains even powers of $1/z$ in addition to the series of odd powers which is the characteristic of real E_O. Again assume that $\mathbf{E}_{OR} = \mathbf{H}_{OR} = 0$ in equation (8.10).

Therefore let

$$E_O = \bar{E}_O \exp[-ikf(A)], \tag{8.27}$$

where $\bar{E}_O$ is the amplitude of the wave E_O and $f(A)$ is the phase deviation from the plane $(x, y, 0)$ for the transmitted wave-front or the deviation of the effective receiving surface from the plane (x, y, z). $f(A)$ is a function of the horn apertures only and is here expressed as a distance.

In practice, it is only necessary to treat phase distortions which, although they may be large in extremes, change symmetrically and gradually over the aperture surfaces; no poorer phase form would ever be used for precision c_0 determinations.

D_z can then still be calculated with sufficient accuracy from equations (8.20), (8.23), (8.24) and (8.27). This is because $\boldsymbol{\phi} \times \bar{E}_O$ and $\boldsymbol{\phi} \times \bar{H}_O$ deviate numerically from E_O and H_O by an amount proportional to the cosine of the angle between the Poynting vector of the radiated energy and the normal to the aperture plane when horns are used as transmitters. Thus the effect of the real part of the phase distortions is to modify E_O by a factor proportional to this cosine, manifesting itself upon the least-squares evaluation of the results as a negligibly small reduction in the effective size of the apertures.

Case 3. $E_O = E_{ax} \cos(\pi x/2a) \exp[-ik(Px^2+Qy^2)]$. Here E_O has the theoretical form for the TE_{01} propagation mode and $f(A)$ can be fitted to a number of phase distortions which are symmetrical and gradual by judicious choice of the constants P, Q. (For phase distortions spherical in form, $P = Q$.) Then

$$D_z = (1/k)[\lambda \beta_1/z - \lambda^2 \beta_2/z^2 - \lambda^3 \beta_3/z^3 + 0(1/z^4) + \ldots], \qquad (8.28)$$

where

$\beta_1 = 4(X_1 + U_1)$.

$\beta_2 = (2/\pi)(Y_1 + V_1) + 4(Y_2 + V_2 + 2X_1 Y_1 + 2U_1 V_1)$,

$\beta_3 = (2/\pi)[3X_2 - 5X_1^2 - 5Y_1^2) + (3U_2 - 5U_1^2 - 5V_1^2) +$

$+ 4(X_1 U_1 - Y_1 V_1)] + 8[(X_3/3 + 3X_1 X_2 - 3Y_1 Y_2 + 10X_1 Y_1^2 -$

$- 10 X_1^3/3] + (U_3/3 + 3U_1 U_2 - 3V_1 V_2 + 10 U_1 V_1^2 - 10 U_1^3/3)]$,

with

$X_1 = a_1^2 (0.2976 - 0.0227 p^2 - 0.0004 p^4)$,

$X_2 = a_1^4 (0.1942 - 0.0308 p^2 - 0.0002 p^4)$,

$X_3 = a_1^6 (0.1652 - 0.0380 p^2 - 0.0001 p^4)$;

$Y_1 = a_1^2 (0.1057 p - 0.0020 p^3 - 0.0002 p^5)$,

$Y_2 = a_1^4 (0.1034 p - 0.0043 p^3 - 0.0002 p^5)$;

$U_1 = b_1^2 (0.5235 - 0.0328 q^2 - 0.0033 q^4 - 0.0003 q^6)$,

$U_2 = b_1^4 (0.4935 - 0.0618 q^2 - 0.0052 q^4 - 0.0003 q^6)$,

$U_3 = b_1^6 (0.5537 - 0.0957 q^3 - 0.0073 q^4 - 0.0004 q^6)$;

$V_1 = b_1^2 (0.2193 q + 0.0069 q^3)$,

$V_2 = b_1^4 (0.2953 q + 0.0059 q^2 - 0.0003 q^5 - 0.0001 q^7)$;

and

$$p = 2kPa^2/\pi, \quad q = 2kQb^2/\pi,$$
$$a_1 = a/\lambda, \qquad b_1 = b/\lambda.$$

It is now necessary to discuss the probable effect of E_{OR} and H_{OR}. This reflected wave may exist either just inside the horn aperture (the most probable

8. MICROWAVE AND RADIO-WAVE INTERFEROMETER METHODS 109

hypothesis) or immediately outside the boundary of the mouth. In this latter instance E_{OR} would be zero and radiation would arise only from an electric current sheet. However, in the conditions of operation of the interferometer the final result of all reflected waves can be attributed to their being obtained from a sum of hypothetical reflected waves E_{OR} equivalent to a single wave representing an integrated effect. It is also satisfactory to regard E_{OR} as existing at the boundaries of the horn-mouths.

First suppose that E_{OR} is either exactly in or out of phase with E_O. The effect of E_{OR} for case 1 and 2 is then to alter slightly the aperture sizes, as given by the least-squares solutions, away from the purely theoretical sizes obtained by neglecting E_{OR}. (The direction of polarization of E_{OR} is always unimportant).

Secondly, suppose there exists a phase difference between E_{OR} and E_O: the worst possible difference is $\pi/2$ and then, for cases 1 and 2, the apparent aperture sizes are virtually the same as the geometrical, but the series for D_z contains even powers of $1/z$. Case 3 behaves differently, and if $f(A)$ averages about $\lambda/4$ between the aperture centres and edges then the effect of E_{OR} is reversed and the apparent aperture sizes are altered most if the phase of E_{OR} is $\pi/2$ relative to the peripheral phase of E_O. Thus the experiments can be used to estimate the amplitude and phase of E_{OR}.

Experiments with the 24 GHz prototype 4-horn interferometer (Froome 1954) were undertaken to test the behaviour for the plane-wave case 2 and spherical transmitted wavefront, case 3. Excellent agreement with theory was obtained, so that the residual uncertainty in the elimination of the diffraction correction for the 72 GHz experiment could be assessed with accuracy and confidence. The millimetre-wave interferometer was always used with phase-correcting lenses fitted to the horn apertures so that the diffraction effect was eliminated by the use of the case 2 formula, as the analysis showed E_{OR} to be in anti-phase to E_O.

Table V gives the measured phase velocities for the seven different positions of the transmitting horns reduced to vacuum conditions.

TABLE V

Transmitter to receiving horn separation (Movable carriage in central position) (cm)	Measured phase velocity (km/sec)
629·5	299,796·020
751·5	4·981
875·0	4·283
999·0	3·802
1120·5	3·583
1247·5	3·482
1367·5	3·259

Table VI gives the corresponding values of c_0 after removal of the diffraction correction.

It is seen that the standard deviation of a single determination of velocity at any particular value of the horn separation is just under 2 parts in 10^7. It should be noted that the variation of the c_0 results in Table VI includes, in addition to the uncertainty of setting on a minimum, the random uncertainties of the air refractometer and the temperatures of the length-bar and the air.

As with the other microwave interferometers (Froome 1952, 1954) it was found that the actual diffraction correction was less than the purely theoretical value. That is, insertion of the geometrical horn apertures (8×6 cm) gave a value of K less than unity when the results were evaluated although the value of c_0 obtained was the same as that finally accepted. The values of c_0 in Table II have been derived by assuming dimensions of $7 \cdot 2 \times 5 \cdot 2$ cm.

Table VII lists the estimates of the possible systematic errors present in the determination.

TABLE VI

Final results

Horn separation (cm)	c_0 (km/sec)
629·5	299,792·513
751·5	2·529
875·0	2·476
999·0	2·414
1120·5	2·478
1247·5	2·588
1367·5	2·512

mean $c_0 = 299{,}792{\cdot}501 \pm 0{\cdot}059$ [s.d.]

TABLE VII

Systematic errors

source of error	magnitude expressed (km/sec)
Length measurement (± 2 in 10^7)	$\pm 0{\cdot}060$
Refractive index of air ($\pm 1{\cdot}1$ in 10^7)	$\pm 0{\cdot}033$
Length-bar temperature ($\pm 0{\cdot}006$ °C)	$\pm 0{\cdot}020$
Air temperature ($\pm 0{\cdot}03$ °C)	$\pm 0{\cdot}010$
c.p.i. residual	$\pm 0{\cdot}015$
Diffraction residual	$\pm 0{\cdot}010$
Frequency (± 1 in 10^8)	$\pm 0{\cdot}003$

The standard deviation of a single determination (Table VI) is increased to 0·10 km/sec when modified by statistical combination with the estimated systematic errors, so the result for the 72 GHz interferometer was

$$c_0 = 299{,}792{\cdot}50 \pm 0{\cdot}10 \text{ km/sec}.$$

It is of interest to give at this point the result of the investigation with the 24 GHz prototype four-horn interferometer, (12·5 mm wavelength) which was first published as $c_0 = 299793{\cdot}0 \pm 0{\cdot}3$ km/sec. Subsequently (Froome 1958) an error was found in the length standard used, so the correct value was

$$c_0 = 299{,}792{\cdot}75 \pm 0{\cdot}30 \text{ km/sec}.$$

The fact that the final value obtained by the millimetre wave interferometer at the NPL is the same as that given earlier by the cavity resonator method is to some extent fortuitous in view of the ten-fold increase in precision.

4. FLORMAN'S RADIO-FREQUENCY INTERFEROMETER

This method consisted of a radio-wave interferometer operating in the V.H.F. range of frequencies at 172·8 MHz (Florman 1955). The experiment was undertaken on a dry lake bed in Arizona, the nearest reflecting objects being at least 7 km from the site. It was considered that the V.H.F. wavelength chosen was short enough to be free from sky-wave interference and from ground effects. The value of the distance requiring precision measurement was of the order of 1 km.

Figure 8.6 shows in diagrammatic form the basis of the method. The baseline is of length D, separating two radio receivers R_1 and R_2. Two values of D were used; 850 m and 1500 m, corresponding to 490 and 864 V.H.F. wavelengths respectively. D was in each case measured by means of 50 m invar tapes to an estimated accuracy of 2 parts in 10^6. T_1 represents the principal V.H.F. transmitter operating at a frequency f_1, of 172·800 MHz. The proportional accuracy of frequency measurement was much higher than that of the length measurement.

T_1 was separated from R_1 by a distance d_1, but the difference in phase between the signals at R_1 and at R_2 will be just $2\pi D/\lambda_1$ where λ_1 represents the wavelength corresponding to f_1. To determine this phase difference accurately the following method was employed. A secondary transmitter T_0 was stationed as shown, the frequency f_0 of T_0 being 172·801 MHz. T_0 transmitted signals to R_1 and R_2; at R_1 a beat-frequency signal was produced and then was relayed by U.H.F. transmission to the phase-measuring equipment at P where it has the form

$$\cos 2\pi[(f_0 - f_1)t + d_3/\lambda_0 - d_1/\lambda_1 + \Phi_1].$$

Here, d_3 is the distance from T_0 to R_1 and Φ_1 is the added phase-angle due to the U.H.F. transmission.

Similarly, the signals from T_0 and T_1 were also mixed at R_2 and relayed to P, giving rise to a signal of the form:

$$\cos 2\pi[(f_0 - f_1)t + d_4/\lambda_0 - (d_1 + D)/\lambda_1 + \Phi_2].$$

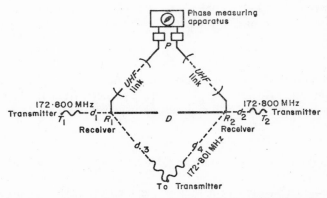

Fig. 8.6. Diagrammatic form of Florman's radio frequency interferometer.

At P, the difference in these phases, $2\pi[D/\lambda_1 + (d_3 - d_4)/\lambda_0 + (\Phi_1 - \Phi_2)]$, was measured. Since only D is known, λ_1 cannot be obtained from this phase comparison alone.

The experiment was then continued as follows: the transmitter T_1 was moved to T_2, still on the line joining R_1 and R_2, but closer to R_2. The phase-difference measured at P was then:

$$2\pi[-D/\lambda_1 + (d_3 - d_4)\lambda_0 + (\Phi_1 - \Phi_2)]$$

which is the same as before except for the change in sign of the term in D. Thus the difference in the two phase comparisons gave $4\pi D/\lambda_1$ and λ_1 could be evaluated from the value of D.

Of course, the phase term $4\pi D/\lambda_1$ contains the whole number of wavelengths in D as well as the fraction of λ_1 which is what is actually obtained by the phase measurement. The whole number was simply obtained by assuming an approximate value for the speed of radio waves.

The determination of the refractive index of air along the transmission paths introduced an estimated additional uncertainty of four parts in a million, largely owing to the great effect of water vapour at radio frequencies.

Florman's final value, based on 110 observations, for the velocity of radio waves was

$$c_0 = 299{,}795 \cdot 1 \pm 3 \cdot 1 \text{ km/sec}.$$

The uncertainty quoted represents a "95% confidence level" combined with an estimate of systematic errors.

5. THE MICROWAVE INTERFEROMETER OF SIMKIN, ET AL.

Recently, Simkin et al. (1967) have repeated the method of Froome (1952) to make a measurement of the velocity of microwaves with his moving-mirror type of interferometer. Their principle innovations were to use a somewhat higher microwave frequency (37 GHz, corresponding to a wavelength of 8 mm) and a larger mirror displacement, namely 447 cms. This distance was obtained by multiplication from two fused quartz end-gauges each 10 cm in length and measured by optical interferometry. The microwave refractive index of the air in the neighbourhood of the instrument was determined by means of a cavity resonator refractometer. The value obtained for the speed of microwaves was:

$$c_0 = 299{,}792.56 \pm 0.11 \text{ km/sec}.$$

The estimated uncertainty is stated to be an 'rms' value.

CHAPTER 9

Modulated Light Beam Methods of High Precision

1. INTRODUCTION

The work of Bergstrand (at the Geographical Survey Office, Stockholm) is outstanding amongst the optical determinations of the velocity of light. With great ingenuity he improved the Kerr cell method of Hüttel to the standard required of a commercial geodetic measuring instrument which is now manufactured by AGA of Sweden under the trade-name of "Geodimeter". The idea of utilizing a modulated light beam for the measurement of geodetic distances between fixed survey pillars was conceived in 1943. In the process of the development Bergstrand has published (over the period 1949 –57) a number of velocity determinations, all of high accuracy.

The Geodimeter represents one of the great advances in the field of long distance metrology. Together with the microwave instrument known as the Tellurometer (described in Chapter 10) it has been responsible for a revolution in surveying. The old method of triangulation from base-lines laboriously measured by invar tapes or wires, has given way to trilateration —the measurement of triangle sides. Not only has the accuracy of surveying been improved, but the increase in speed is probably even more important. Of course, an accurate value for the speed of light was essential if trilateration was to become a reality, and Bergstrand's first value was presented to the International Union of Geodesy and Geophysics at Oslo in 1948 and was published in 1949. (Bergstrand 1949a). Thus it was annnounced a year or so after Essen and Gordon—Smith's microwave cavity resonator measurement had thrown doubt on Birges (1941) statistically recommended figure.

Another light-modulator described very recently is that of Karolus and Helmberger which utilizes the properties of ultrasonic standing waves in a liquid. A value of the speed of light has been obtained from this new method.

2. BERGSTRAND'S GEODIMETER

Figure 9.1 shows in diagrammatic form the principles of the method (Bergstrand 1949b). A crystal controlled oscillator of approximately 8 MHz was used to drive a Kerr cell light modulator K, the output of which

9. MODULATED LIGHT BEAM METHODS OF HIGH PRECISION

was focussed by means of a concave reflector into a parallel beam directed at the target reflector R. The light source was a tungsten filament lamp. The reflected light from the target (placed at the far end of the known distance) was detected by a photo-multiplier, PX, placed at the focus of a second concave mirror, about 45 cm diameter, identical to the transmitter. The dynodes of the photo-electric cell were energized by a high tension alternating potential also derived from the 8 MHz quartz oscillator.

M represents the meter which indicated the photo-tube anode current, and if D is taken as the distance from the instrument to the far reflector, the photo-current varied in the manner of Fig. 9.1a as D was increased by moving the distant reflector. One cycle of this variation represented a half-wave of the modulation wavelength corresponding to 8 MHz. Figure 9.1b shows the similar variation detected when the phase of the modulating electric field applied to the Kerr cell was changed by 180°.

The greatest sensitivity is clearly to be attained by combining these curves as shown in Fig. 9.1c, which represents the output of the null meter employed in practice. To achieve this mode of operation, Bergstrand additionally biased the Kerr cell with a 50 Hz high voltage square wave so as to produce a sinusoidal modulation at 8 MHz, the phase of which was reversed 100 times a second. Figure 9.2a shows how the light output from the Kerr cell varied with the total applied potential across the plates. The 50 Hz bias was adjusted so that the 8 MHz field oscillated between the points a and b giving rise to a reasonably sinusoidal 8 MHz intensity modulation superimposed on a "d.c." background. The 50 Hz generator also controlled the electronic synchronous "switch" shown in Fig. 9.2b which was placed in the output from the photo-multiplier. It is easy to see how this switch sums the curves of Figs 9.1a and 9.1b to produce 9.1c, as each valve in it conducts during alternate halves of the 50 Hz biasing voltage. Thus the meter M would read a null for every quarter-wave displacement of the far reflector. In practice this switching process is not perfectly symmetrical, neither is the behaviour of the Kerr cell when the d.c. bias is reversed. To eliminate these effects two manually operated change-over switches (not shown) were provided which reversed the 50 Hz and also the high frequency phases. The measuring process with all Geodimeters involves the use of these manual switches.

Figure 9.3 shows the Geodimeter system in detail. S is the source of light, focussed down by means of the lenses shown to form a small image between the plates of the Kerr cell, K; P_1 and P_2 are Nicol polarizing prisms, crossed relative to each other, the plane of polarization from P_1, being at 45° to the electric field in the nitrobenzene-filled Kerr cell. The main transmitting optics were focussed also at this point in the Kerr cell. C represents the quartz-crystal controlled oscillator which supplied to the Kerr cell a high potential of about 2000 V alternating at a frequency of 8 MHz. A 50 Hz square-wave generator switched the bias polarity of 5000 V on the

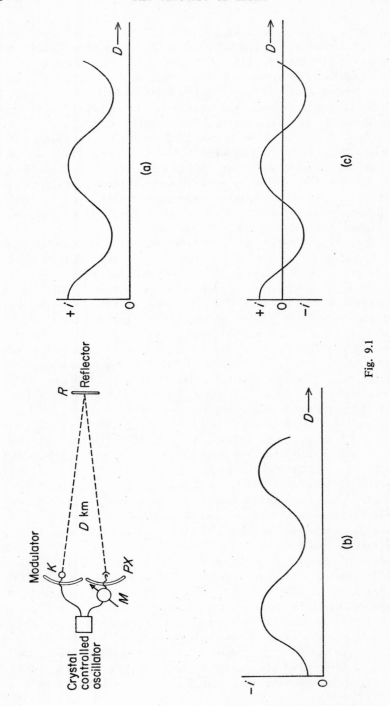

Fig. 9.1

9 MODULATED LIGHT BEAM METHODS OF HIGH PRECISION

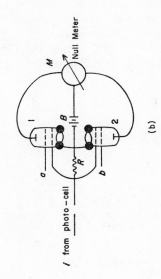

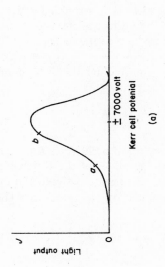

Fig. 9.2

Kerr cell at 100 times a second, and also operated the synchronous switch depicted in Fig. 9.2b. Following the 8 MHz oscillator was placed an electrical delay line, *DL*, the purpose of which was to vary the phase of the r.f. signal applied to the photo-multiplier PX. When the Geodimeter is in use this variable delay is operated until the null indicator reads zero. For calibration

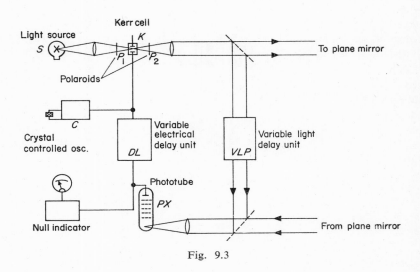

Fig. 9.3

purposes a variable light-path (VLP) was provided which could be placed in front of the instrument and used to transfer with variable modulation phase a portion of the transmitted light to the receiving system.

The theory of the method is as follows:

The intensity of light, *J*, transmitted by the second nicol prism N_2 is given by

$$J = J_0 \sin^2 K.V^2, \qquad (9.1)$$

where J_0 is the value of *J* for *uncrossed* polarizers, *K* is a constant and *V* is the potential difference between the plates of the Kerr cell.

By biasing the Kerr cell so as to operate between the points *a* and *b* of Fig. 9.2 a sinusoidal 8 MHz modulation was obtained. The voltage across the cell is, say,

$$V = V_0 \sin \omega t, \qquad (9.2)$$

where V_0 is the amplitude of *V* and $\omega = 2\pi f$.

9. MODULATED LIGHT BEAM METHODS OF HIGH PRECISION

There will be a small time delay t_1, between the modulating voltage and the modulated light intensity transmitted by N_2. This modulated light will have the form:

$$J_{50+} = C_1 + C_2 \sin \omega(t - t_1) \tag{9.3}$$

$$J_{50-} = C_1 + C_2 \sin [\omega(t - t_1) + \pi] \tag{9.4}$$

where $50+$ and $50-$ indicates the light transmitted during the positive or negative half-cycles of the 50 Hz square-wave bias on the Kerr cell. C_1 is the constant part of the transmitted light intensity and C_2 the amplitude of the variations.

After traversing the distance D to the far reflector R, this light is focussed by the receiving telescope system on to the photo-multiplier PX. This tube is supplied with 8 MHz oscillations, but *not* with the 50 Hz square-wave which in fact operates the synchronous electronic switch between the photo-tube and the null meter M. Current flows in the photomultiplier only during the positive half-cycles of the high frequency. Assuming rapid saturation the photo-current is:

$$i_{50+} = \frac{f}{2} \int_{t_3}^{t_3+1/2f} [A + B \sin \omega(t - t_1 - t_2 - 2D/c)] \, dt \tag{9.5}$$

$$i_{50-} = \frac{f}{2} \int_{t_3}^{t_3+1/2f} [A + B \sin \omega(t - t_1 - t_2 - 2D/c) + \pi] \, dt \tag{9.6}$$

A and B correspond to C_1 and C_2 of the previous equations and c is the group velocity of light for the prevailing atmospheric conditions. t_2 is the transit-time of the electrons in the photo-multiplier. The summation is for one positive half-cycle of the 8 MHz modulation which begins at t_3 and ends at $t_3 + 1/2f$. The current (charge/sec) is obtained by multiplying by $f/2$ the number of half-cycles in one second of $50+$ or $50-$ biasing. The synchronous electronic switch of Fig. 9.2 ensures that the two currents i_{50+} and i_{50-} have opposite effect on the null meter M. Thus the meter current is i, where

$$i = i_{50+} - i_{50-} = \frac{B}{\pi} \cos \omega(t_3 - t_1 - t_2 - 2D/c) . \tag{9.7}$$

When the meter reads zero

$$\omega(t_3 - t_1 - t_2 - 2D/c) = \pi/2 - N\pi , \tag{9.8}$$

where N denotes a whole number. Thus

$$D = \frac{f}{2}(t_3 - t_1 - t_2) \lambda + \frac{2N-1}{8} \lambda \tag{9.9}$$

or

$$D = K + \frac{2N-1}{8} \lambda \tag{9.10}$$

where K is a constant which can be eliminated by measuring two values of D. The difference D_N, is given by

$$D_N = \frac{2N-1}{8}\lambda. \qquad (9.11)$$

From this equation λ is found and thus

$$c = f\lambda. \qquad (9.12)$$

In equation (9.11), D_N represents the length of an accurately known baseline. Bergstrand's velocity determinations did not rely upon the accuracy of the internal electrical delay line or the external variable light path, although some other Geodimeter measurements to be discussed have done so.

3. GEODIMETER RESULTS FOR c_0

Bergstrand's preliminary determination was made at the Linkoping base in Sweden using the full length of 9·07 km and the "half" value of 4·2 km. The measurement was made entirely by displacing the targetreflector from a station near the Geodimeter to the far end of the base. Linkoping was not considered the most accurate base available, Bergstrand, himself estimating it with an uncertainty of 2 parts in 10^6 His first published result for the velocity of light (Bergstrand 1949a) was

$$c_0 = 299,796 \pm 2 \text{ km/sec}.$$

The error-limit quoted represented the mean instrumental reading error in statistical combination with the estimated uncertainty of the base together with additional uncertainties caused by the effects listed in Table II. In fact, Bergstrand subsequently considered that this value was too high owing to the disturbing effect on the photo-multiplier tube behaviour of the very powerful (and large image-size) signal returned from the reflector in the near position. In his other determinations a convex reflector was used at the near-point, its radius so chosen to produce substantially the same image size and intensity as that produced from a plane mirror or corner reflector at the far point. In conjunction with the switches for the additional manual reversing of the 50 Hz and 8 MHz phases (already referred to) this improvement in experimental technique greatly reduced the inherent instrumental errors.

Bergstrand's (Bergstrand 1950, 1951) principal determinations were obtained from two well established bases, one of 6·9 km and the other of 5·4 km. The actual experimental work was undertaken in May to September 1949 at 6·9 km, and August 1950 at 5·4 km.

9. MODULATED LIGHT BEAM METHODS OF HIGH PRECISION

Table I shows the results obtained.

Each determination is the mean of many working nights, each single night's observations being averaged in a preliminary reduction. The result of the 1951 determination on the Öland base of 5·413 km was

$$c_0 = 299{,}793 \cdot 14 \pm 0 \cdot 42 \text{ km/sec}.$$

Table II shows how Bergstrand assessed his errors.

Geodimeters use essentially "white" light sources so that the "colour uncertainty" arises from uncertainty in the assessment of the effective operating wavelength.

It is seen that an extremely small and possibly optimistic error (3×10^{-7}) was quoted for the invar-wire determinations of the bases used. Atmospheric refractivity errors can be assessed from the fact that a change of 1×10^{-6} is produced by a change of $1°$ C in air temperature or 2·6 mm Hg in pressure. Humidity has a very small effect in the visible region of the spectrum.

The base of 5·4 km at Öland was used for another careful determination by Schöldström (Schöldström 1955) using a commercial model Geodimeter. His result was

$$c_0 = 299{,}792 \cdot 4 \pm 0 \cdot 4 \text{ km/sec}.$$

TABLE I

Year	Base (km)	c_0 (km/sec)
1949 May	6·906	299,793·04 ± 0·19
1949 September	6·906	3·10 ± 0·20
Final mean (1950)		3·1 ± 0·26

TABLE II

Nature of error	Magnitude (m/sec)
Random (Deviation from mean)	± 150
Refractive index formula	± 60
Colour uncertainty	± 60
Atmospheric measurements of pressure and temperature	± 60
Base-line	± 90
Frequency	± 60
Statistical mean	± 220 m/sec

Commercial Geodimeters were also used by Mackenzie (Mackenzie 1954) on the two principal British bases at Ridge Way (11·3 km) in Wiltshire and Caithness (24·8 km) in Scotland. From these observations Edge (Edge 1956) obtains values for c_0 of

$$c_0 = 299{,}792 \cdot 4 \pm 0 \cdot 11 \text{ km/sec (Ridge Way Base)}$$

$$c_0 = 299{,}792 \cdot 2 \pm 0 \cdot 13 \text{ km/sec (Caithness Base)}.$$

The limits quoted here represent RMS errors only. The bases were measured by 24 m invar tapes in catenary to an accuracy which can be estimated at about $\pm 1 \times 10^{-6}$ for Caithness and $0 \cdot 7 \times 10^{-6}$ for Ridge Way. (Allowance was made for the stated internal consistency plus tape calibration errors, see also Chapter 10.) If one also makes allowance for uncertainties in the measurement of atmospheric pressure and temperature the limits of error would need to be considerably increased.

In 1957, Bergstrand (Bergstrand 1957) published an evaluation of all the Geodimeter determinations of c_0 up to that time. Table III gives what he calls "compensated" measurements in which the velocity measurement was not subject to possible inherent instrumental errors and was obtained from positioning of the target-reflector at the near and far-points corresponding to each end of the base-line. He has assumed an error-limit of $\pm 0 \cdot 38$ km/sec on c_0 for each determination listed.

The total weighted mean result from Table III

$$c_0 = 299{,}792 \cdot 85 \pm 0 \cdot 16 \text{ km/sec}.$$

Table IV gives "uncompensated" results with early Geodimeters.

TABLE III

Base-line	Distance (m)	Velocity (km/sec)	Mean of base	Weight
Swedish	6906	299,793·02	3·05	10
	6906	3·08		
	5413	3·17	2·80	8
	5413	2·43		
	7320	3·37	3·37	7
Australian	6440	2·65	2·44	11
(see Bergstrand 1957)	6440	2·05		
	6440	2·64		
	9660	2·41	2·46	14
	9660	2·50		
	6440	3·61*	3·61	3
	11100	3·21*	3·21	6

* These strictly are "uncompensated" values but made with improved Geodimeters

9. MODULATED LIGHT BEAM METHODS OF HIGH PRECISION

The weighted mean from Table IV is

$$c_0 = 299{,}792 \cdot 75 \pm 0 \cdot 34 \text{ km/sec}.$$

It is seen that the average of these values is only 1×10^{-6}, higher than the most precise microwave value (Chapter 8).

Fig. 9.4a

Figure 9.4a is a photograph of the earliest Geodimeter. Fig 9·4b is the latest model (MK6) Geodimeter shown diagrammatically in Fig. 9.5. It is seen that the optical system has evolved from separated transmitting and receiving optics into a co-axial system with the receiving telescope inside the transmitted beam of light. Otherwise, all the features of the earlier

TABLE IV

Base line	Distance (m)	Velocity (km/sec)	Weight
Australian	6440	299,792·50	6
English	11260	2·40	11
	24830	2·20	25
USA	1380	4·06	1
(See Bergstrand 1957)	12800	4·27	13
	3120	2·73	3
	2130	1·69	2

instruments are preserved, a calibrated variable electrical delay line being used to measure the phase of the light-modulation returned from the far reflector by adjusting it until the null meter reads zero. The modulation

Fig. 9.4b

frequency has been increased to 30 MHz to obtain improved resolution which is claimed to be 1 cm on the shorter ranges. In daylight, the maximum range is 3 km with a tungsten lamp or 6 km with a mercury-vapour lamp. The corresponding darkness ranges are 15 km and 25 km using clusters of cube corner reflectors at the far station.

The required distance is obtained by first obtaining the fraction of a modulation quarter-wave (2·5 m) in it and then obtaining the integer number of quarter-waves by changing the modulation frequency by a small proportion in steps and re-measuring the phase each time. This is analogous to the "method of fractions" used in optical interferometry to obtain the order.

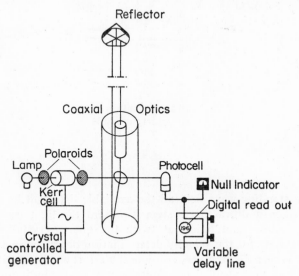

Fig. 9.5

Essentially the same procedure is used in the "Mekometer" method of distance measurement outlined in Chapter 11 and the "Tellurometer" method described in Chapter 10.

4. THE ULTRASONIC LIGHT-MODULATOR

Karolus and Helmberger (1964) have developed a modulator which uses the diffraction of light passing through an ultrasonic standing wave system generated in carbon-tetrachloride, paraffin or silicon oil, by quartz crystal oscillators immersed in the liquid. The velocity of sound in the liquids investigated is of the order of 1 km/sec. so that the spacing of the wave-system was in the region of 0.1 mm for driving frequencies 10 MHz.

Figure 9·6 shows the basis of the modulator. Monochromatic light is required and this was obtained by use of an interference filter placed between the mercury-vapour light source L and first slit S_1. For transmission through the ultrasonic cell the light was rendered parallel by the lens L_1, the lens L_2 focussing the resulting Fraunhofer diffraction pattern on to the slit S_2. The most successful standing wave generator employed two synchronously-driven quartz crystals Q_1 and Q_2, mounted as shown. It is seen that the

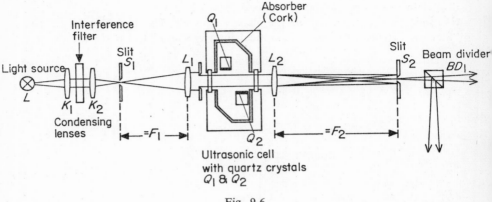

Fig. 9.6

absence of a standing-wave field all the available light will be transmitted by S_2, but when a diffracting system is present the light leaving S_2 will undergo a shallow modulation at twice the quartz crystal frequencies, namely at 19 MHz for the velocity determination described in 1966 (Karoius and Helmberger 1966). A very considerable effort was involved in producing a modulator with a sufficiently uniformly phased standing-wave system (Karolus and Helmberger 1964).

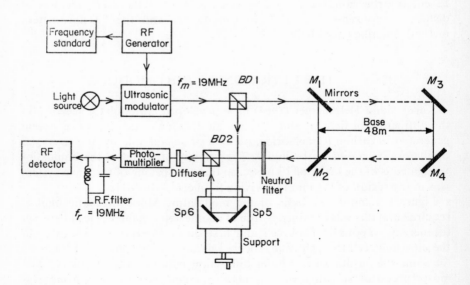

Fig. 9.7

9. MODULATED LIGHT BEAM METHODS OF HIGH PRECISION

Figure 9·7 shows the complete experimental layout for the measurement of the velocity of light. Following the slit S_2 (of Fig. 9·6), a beam-divider ($BD1$) was used to provide a reference signal for the photo-multiplier cathode and the phase of this signal could be varied by means of a micrometer-operated variable light path. The "measuring light" was transmitted through the beam-divider ($BD1$) to the base-line of 48 m. A double-mirror system was used to return this light from either end of the base. A second beam-divider ($BD2$) combined the modulated light from the base-line reflectors with that from the variable light path before entering the photo-multiplier. The output from the photo-tube at 19 MHz was detected by a suitable filter and meter system.

It is seen that if the modulated light reflected from the base arrives at the photo-cathode in phase with that from the variable light path, the 19 MHz component of the photo-multiplier will be a maximum. It will be a minimum when the light beams arrive in anti-phase. In order to attain a high sensitivity of modulation phase setting, a very careful balance of the two intensities was essential, so that an adjustable neutral filter was placed in the light-path from the base line in order to make this balance. By this means movements of the variable light path of less than 0·1 mm could be detected, so this represented the precision of determination of the base-line end-points. The method of wavelength-measurement was to set to a minimum of the 19 MHz photo-tube output.

Because the base-line was short it could be measured with precision by the method of Vaisala (1958). This is a method employing optical interferometry and is described in Chapter 11. The estimated error was $\pm 1 \times 10^{-7}$.

The residual uncertainty arising from the correction for atmospheric refractive index were low owing to the simplicity of the pressure, temperature and humidity measurement. The error of the refractive index correction was assessed at $\pm 1 \times 10^{-7}$.

The final result for the velocity of light was $c_0 = 299{,}792 \cdot 1 \pm 0 \cdot 2$ km/sec. Subsequently an error in the base length was reported (Baird 1967) and the value was adjusted to

$$c_0 = 299{,}792 \cdot 44 \pm 0 \cdot 2 \text{ km/sec}.$$

The error-limit quoted is obtained from approximately 14000 observations averaged in 278 groups of 50 measurements. The subsequent correction to the value assigned for the 48 m base-line emphasizes the very real difficulties that can be encountered in precision length measurement even over these relatively short distances.

CHAPTER 10

The Tellurometer

1. INTRODUCTION

Although accurate determinations of the velocity of microwaves have been obtained from Tellurometer measurements, in particular at the British Ordnance Survey's base at Ridge Way, the greatest significance of the Tellurometer is in the field of geodetic distance measurement. The instrument was invented by Wadley of the Commonwealth Scientific and Industrial Research Organisation (South Africa) following an assessment in 1954 of the need for such an instrument by Colonel Bauman, the Director of Trigonometrical Survey in the Union of South Africa. The suggestion was to produce a microwave system with a range of 20 miles or more which would be accurate to better than 1 in 10^5 and would work in daylight and haze when the optical modulator methods fail through loss of signal.

In 1956 the Tellurometer Company (South Africa) was established to develop the instrument commercially under licence from the government. At about this time a world market survey was instigated to assess the probable demand, which was estimated to be 250 instruments. By 1967 over 6000 Tellurometers had been sold, the large demand arising from the enthusiasm with which surveyors adopted it for the measurement of the triangle sides, instead of persisting with the traditional method of measuring angles from one or two carefully measured and maintained geodetic base-lines.

The Tellurometer differs from the Geodimeter in that it uses a microwave carrier radiation. This is modulated at 10 MHz, and transmitted from a small parabolic antenna to a far station which compares its phase with that of an internal oscillator and then returns this phase information to the master station. Thus the phase of the total signal at the master received from the remote station behaves as though the 10 MHz modulation was reflected back passively with a small fixed phase error. Transmitter and far-station in fact are substantially identical. Wadley's original version operated with a 10 cm (3 GHz) wave-length radiation, but later models use 3 cm (10 GHz). Recently (1967) the Tellurometer Company have announced a high resolution instrument which has an 8·6 mm (35 GHz) carrier wavelength.

2. THE METHOD OF MEASUREMENT

The Tellurometer was first described in 1957 (Wadley 1957). The results of velocity determinations were given in the following year (Wadley 1958a) and the electronic principles were described in detail in a separate paper (Wadl y 1958b).

The original model was intended to measure in feet and employed as basic standards quartz crystal oscillators of frequency near 10 MHz. The microwave carrier radiation was obtained from low-power klystron oscillators which were frequency- modulated by the standards. The disturbing effect on modulation phase measurement of reflections from the ground of the microwave beam had to be eliminated and this was achieved by varying the klystron frequency by the amount required to cycle this effect positively and negatively relative to the mean (undisturbed) state. All Tellurometers involve the use of this technique.

Figure 10.1 is a diagram of the Tellurometer system. The primary modulation frequency of 10 MHz at the master station gives rise, after conversion at the remote relay station, to a phase pattern (A) repeating itself at roughly 50 ft intervals, the final precision of phase determination being approximately 1° or 1·5 in on length. The difference in phase readings between this modulation pattern and the 9·000 MHz (D-pattern) phase readings gives rise to a pattern which repeats every 500 ft. Similar $A - C$ phases repeats every 5000 ft and $A - B$ every 50,000 ft. Thus the fraction of 50 ft in the required distance is obtained first and then by successive approximation the integer number of 50 ft units is also obtained. Much the same basic method for obtaining this "order number" is used in the "Geodimeter" (Chapter 9) and "Mekometer" (Chapter 11).

Although the distance is effectively measured by the time or phase delay of a 10 MHz wave after having made the double journey, the technical requirements are met by an ingenious modification which need be explained for the A-pattern only. The outgoing modulation signal at 10 MHz is combined with a modulation signal of 9·999 MHz at the remote station, thus giving a 1 kHz difference frequency, the phase of which is governed by that of the received 10 MHz signal. After amplification the 1 kHz signal is applied as a frequency modulation to the transmitted signal at 9·999 MHz. When this signal is received at the master station it will again give a difference frequency at 1 KHz which will carry the phase lag of the 9·999 MHz wave in addition to that of the 1 KHz modulation. The receivers at the two stations are, of course, of the heterodyne type and have an intermediate frequency of 33 MHz. The two carrier frequencies differ by 33 MHz so that each acts as the local oscillator with which the received signal is mixed. Let

f_m be the master modulation frequency (10 MHz),
f_r be the remote modulation frequency (9·999 MHz).

130 THE VELOCITY OF LIGHT

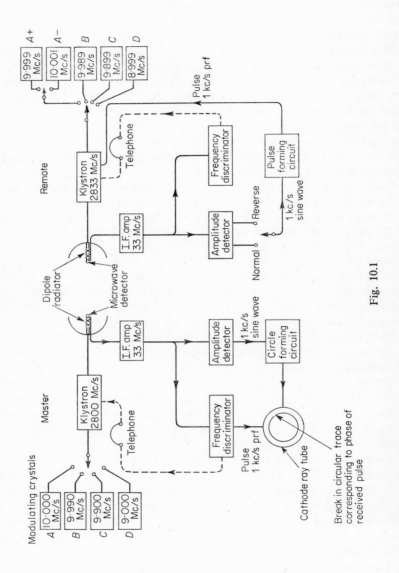

Fig. 10.1

10. THE TELLUROMETER

f_c be the comparison frequency (1 KHz),
$T/2$ by the delay time of the signal in one traverse of the distance.

Then the phase change due to the forward path alone is $2\ \pi f_m T/2$ and that due to the return path is $2\ \pi f_r T/2$ to which is added the phase change of the 1 KHz modulation of $2\ \pi f_c T/2$. The total phase change at the master station is thus $\pi T(f_m + f_r + f_c)$ and as $f_c = f_m - f_r$ the final phase change is $2\ \pi f_m T$, which is that which would be obtained of the master wave at f_m had simply been reflected at the remote station. The advantages of imposing the phase changes on to the 1 kHz modulation are that the unavoidable phase shifts in the receiving equipment can be made small and constant, and by using a transmitter at the remote station the sensitivity is enormously increased. The residual phase errors in the system were eliminated by Wadley in the following way. The remote frequency f_r could be changed from 9·999 MHz to 10·001 MHz so that the change of the phase of the comparison frequencies derived from the primary modulations with increasing separation of the stations would be reversed. However the indicated phase due to the comparison frequency path delay alone remains the same. The expression for the total indicated phase change becomes, for the A-pattern $\pi T[-f_m - f_r + f_c] = -2\ \pi f_m T$. The phase change is indicated on a circular scale on a cathode-ray tube and if in the former case the indication rotates clockwise with increasing distance it will now rotate anticlockwise. The mean of the phase measurements derived from the "$+A$" and "$-A$" patterns is free from major internal phase errors. Reverse readings were also provided on the positive and negative patterns to eliminate the effect of any eccentricity in the display system. For this purpose the phase of the comparison frequency derived from the remote crystal detector could be reversed by reversing the polarity of the diode. "Negative" patterns were not found necessary for the B, C and D patterns since these are subtracted from the A-pattern and systematic errors are adequately eliminated. In Wadley's first Tellurometer the comparison wave from the remote station was returned as a pulse at a recurrence of 1 kHz and displayed on the cathode ray tube as a break in a circular trace formed from the master 1 kHz comparison wave. The same circuitry also incorporated a two-way radio-telephone.

The mathematical analysis of the first conversion process is as follows. It is assumed that a strong local frequency modulated signal is heterodyned with a received frequency modulated signal. Let the two signals be proportional to:

$$\exp\ [j(\omega t + m \sin \Omega t)]$$

$$[\exp\ [j(\omega^1 t + m^1 \sin \Omega^1 t)]$$

where t is the time, ω and ω^1 the carrier circular frequencies Ω, and Ω^1 the circular modulation frequencies ($\omega > \omega^1$; $\Omega > \Omega^1$). Then the signal at the difference frequency will be proportional to

$$I = \exp[j(\omega t - \omega^1 t + m \sin \Omega t - m^1 \sin \Omega^1 t)]. \qquad (10.1)$$

This may be re-written in terms of Bessel functions, as

$$I = \exp[j(\omega - \omega_1^1)t] \sum_{n=-\infty}^{n=+\infty} J_n(m) \exp(jn\Omega t) \sum_{v=-\infty}^{v=+\infty} J_v(-m^1) \exp(jv\Omega^1 t). \qquad (10.2)$$

Using the identities

$$J_n = (-1)^n J_{-n}$$

and

$$J_n(x) = (-1)^n J_n(-x)$$

and writing

$$\Omega - \Omega^1 = p,$$

I may be evaluated by multiplying term by term the two series resulting from the summations equation in (10.2). Collecting only the terms with frequencies $\omega - \omega^1$, $\omega - \omega^1 \pm p$, $\omega - \omega^1 \pm 2p$, etc. which came within the bandwidth of the *if* amplifier,

$$I = J_0(m) J_0(m^1) \exp[j(\omega - \omega^1)t + J_1(m) J_1(m^1)] \exp[j(\omega - \omega_1 + p)t] +$$
$$+ \exp[j(\omega - \omega^1 - p)t] + J_2(m) J_2(m^1) \cdot$$
$$\cdot \{\exp[j(\omega - \omega^1 + 2p)t] + \exp[j(\omega - \omega^1 - 2p)t]\} + \ldots \qquad (10.3)$$

Equation (10.3) shows that the signal at the intermediate frequency $\omega - \omega^1$ is amplitude modulated at the difference frequency $p = \Omega - \Omega^1$ and its harmonics. The percentage modulation at the fundamental frequency p is $[J_1(m)J_1(m^1)]/[J_0(m)J_0(m^1)]$ and the percentage second harmonic distortion of this modulation is $[J_2(m)J_2(m^1)]/[J_1(m)J_1(m^1)]$. Higher harmonics are negligible for low modulation indices. For $m = m^1 = 1$ one obtains 27% modulation with 6% second harmonic distortion. For $m = m^1 = 0.5$ the values are 7% modulation and 1·3% distortion. The Tellurometer was adjusted to operate between these limits.

3. VELOCITY MEASUREMENTS

Figure 10.2 is a photograph of the first Tellurometer being operated by Wadley. The microwave "dish" was about 45 cm in diameter. Microwave carrier radiation is greatly affected by the high refractive index of water vapour (see Chapter 3, Section 6) at radio and microwave frequencies. At optical frequencies it differs from dry air by only a small amount so that it needs a 20 mm Hg change of water-vapour pressure to make a 1×10^{-6}

10. THE TELLUROMETER

change in refractive index. Thus for distance-measurements with a modulated light-beam, no measurement of humidity is normally made. But for a microwave carrier radiation, the refractive index changes 6×10^{-6} for 1 mm of water

Fig. 10.2

vapour pressure. Thus with the Tellurometer, very careful humidity measurements are essential.

Wadley's first Tellurometer was used in April 1957 to determine the speed of microwaves at the Ridge Way Base of 11·3km in Wiltshire, England. The British Ordnance Survey state that the internal consistency of their measurements by invar-tapes indicates an accuracy of $\pm 0.5 \times 10^{-6}$. The tapes themselves were calibrated at the National Physical Laboratory to approximately the same limits, thereby putting a variation of $\pm 0.7 \times 10^{-6}$ as the overall assessed accuracy of the base-line itself. The refractivity correction was obtained by measurements of atmospheric pressure, temperature and humidity at each end of the base line together with the Essen and Froome (1951) formula given in Chapter 3. In addition to water vapour, an uncertainty of 1° C in temperature corresponds to an 1×10^{-6} error and 2·6 mm Hg of dry air pressure also produces this error. Thus uncertainties in all these measurements give a total error in the refractive index of about 1×10^{-6}.

The value for the Ridge Way Base velocity given by Wadley was

$$c_o = 299{,}792{\cdot}6 \text{ km/sec}.$$

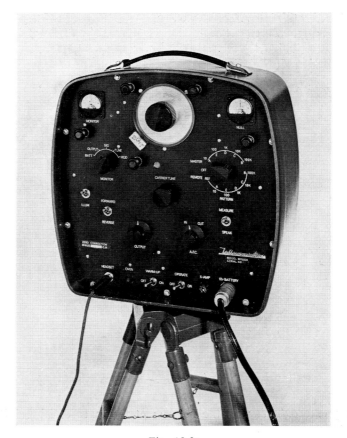

Fig. 10.3a

No specific limits were given, but in the discussion he suggests that the accuracy of this measurement is twice as good as that of his earlier results given below. From these he assessed the accuracy of the Tellurometer as 5 cms $\pm\ 3 \times 10^{-6}$ of the distance.

Wadley had made the two earlier determinations in South Africa, and when he used the Essen and Froome refractivity equation, these became:

Kroonstad velocity: $c_0 = 299{,}792 \cdot 9$ km/sec
Mtubatuba velocity: $c_0 = 299{,}792 \cdot 7$ km/sec.

(In his original paper Wadley used the Smith and Weintraub (1953) refractivity formula which gave results 0·2 km/sec higher in velocity).

Figure 10.3a depicts the MRA 101, probably the most inexpensive and sim-

10. THE TELLUROMETER

Fig. 10.3b

plest Tellurometer produced. With a 10 GHz carrier frequency it has a limiting resolution of about 3 cms. It is in metric units and employs modulation frequencies near 7·5 MHz. The model shown in Fig. 10.3b is the most sophisticated Tellurometer yet to appear. It has a 35 GHz carrier frequency (corresponding to a carrier wavelength of 8·6 mm) and the resolution is much enhanced by the adoption of an additional modulation frequency of 75 MHz. A phase resolution of 3 mm is claimed. The millimeter-wave carrier frequency results in a narrow transmitted beam so that the disturbing influence of ground scatter is much reduced. For centimetric resolution it is usually unnecessary to average through this ground scatter effect by making the customary small changes in carrier frequency.

CHAPTER 11
Summary of Results

The results obtained since 1941 are given in Table I. The precision of many of the measurements has reached a value comparable with that of measuring the path length. For example, the result of Karolus and Helmberger was altered by 1 part in 10^6 because of an error found subsequently in the measurement of the base-line in spite of the fact that it had been determined by Vaisala's method and was thought to be accurate to 0·1 parts in 10^6. There is an additional error due to the uncertainty in the refractive index of the air in the path and this is likely to be greater for radio than for optical waves. When the result depends on the average of many measurements the errors due to refractive index will tend to cancel but that due to the length measurement will remain as a systematic error. The modern methods depending on a base line measurement are of course designed to operate in the reverse sense giving the lengths in terms of the velocity of light. It is no criticism of them therefore to point out that as a means of measuring c they have this fundamental limitation. The methods based on a short distance, such as the cavity resonator method and Froome's radio interferometer, are advantageous from this point of view. In both of these methods the gauges used to determine the change of length were measured directly in terms of the wavelength standard and they could both be elaborated in order to measure the change directly in terms of the wavelength standard. The cavity resonator method has the further advantage of being carried out in a vacuum, but the effect of surface resistance would need to be investigated and eliminated more thoroughly than was done in Essen's early experiment if a higher accuracy is to be obtained. It seems possible that by such methods an accuracy approaching quite closely to that of the wavelength standard itself could be achieved.

TABLE I

Date	Author	Method	Velocity of light c_0 (km/sec)	Spread of results (km/sec)	Given limits of error (km/sec)
1947	Essen and Gordon-Smith	Cavity resonator	299,792	6	± 3
1947	Smith Franklin and Whiting	Radar	299,695*		± 50

* Value in air.

11. SUMMARY OF RESULTS

TABLE I (antinned)

Date	Author	Method	Velocity of ligth c_0 (km/sec)	Spread of results (km/sec)	Given limits of error (km/sec)
1947	Jones	Radar	299,687*		± 25
1949	Aslakson	Radar	299,792·4	11	± 2·4
1949	Bergstrand	Geodimeter	299,796	4·8+	± 2
1949	Jones and Cornford	Radar	299,701*		± 25
1950	Essen	Cavity resonator	299,792·5	2	± 1
1950	Bergstrand	Geodimeter	299,793·1	2+	± 0·26
1950	McKinley	Quartz modulator	299,780	500	± 70
1950	Houstoun	Quartz modulator	299,775	180++	± 9
1950	Hansen and Bol	Cavity resonator	299,789·3		± 0·8
1951	Bergstrand	Geodimeter	299,793·1	5+	± 0·4
1951	Aslakson	Radar	299,794·2	13·5	± 1·4
1951	Froome	Radio interferometer	299,792·6	2·5#	± 0·7
1952	Rank, Ruth and Vanden Sluis	Spectral lines	299,776		± 6
1954	Froome	Radio interferometer	299,793·0	0·7#	± 0·3
		corrected 1958	299,792·75	0·7#	± 0·3
1954	Rank, Shearer and Wiggins	Spectral lines	299,789·8		± 3
1954	Florman	Radio interferometer	299,795·1		± 3·1
1955	Schöldström	Geodimeter	299,792·4		± 0·4
1955	Plyler, Blaine and Cannon	Spectral lines	299,792		± 6
1956	Wadley	Tellurometer	299,792·9		± 2·0
			299,792·7		± 2·0
1956	Rank Bennett and Bennett	Spectral lines	299,791·9		± 2
1956	Edge	Geodimeter	299,792·4		± 0·11**
			299,792·2		± 0·13**
1957	Wadley	Tellurometer	299,792·6		± 1·2
1958	Froome	Radio interferometer	299,792·5	0·17#	± 0·1
1966	Karolus	Modulated light beam	299,792·1	2++	± 0·2
		Corrected 1967	299,792·44	2++	± 0·2
1967	Simkin, Lukin Sikora and Strelenskii		299,792.56		± 0·11++

* Value in air; + Spread of means of a night's observations; ++ Spread of means of 20 observations; †† Spread of means of 50 observations; # Spread after averaging about 10 observations to remove the effect of room reflections. ** S. d. of mean, not including systematic errors.

TABLE II

Basic forms of long distance electronic measuring instruments

Type	Instrument	Range in daylight	Modulation frequency	Stated accuracy	Remarks
Electro optic Kerr cell amplitude modulator	Geodimeter (MK 6) AGA Co.	15 m—3 km (tungsten lamp) 15 m—6 km mercury arc lamp	30 MHz 30 MHz	10 mm $\pm$ 2$\times$10^{-6} 10 mm $\pm$ 2$\times$10^{-6}	Mean square error
Microwave, amplitude modulated	Tellurometer MRA 101 3 cm carrier	50 m—50 km	10 MHz	15 mm $\pm$ 3$\times$10^{-6}	Mean of ten obervations
	MRA 4 Experimental 8 mm carrier	50 m—50 km	75 MHz	10 mm $\pm$ 3$\times$10^{-6} 3 mm $\pm$ 3$\times$10^{-6}	Single observations Mean of 10 observations
Electro-optic quartz modulator	EOS (Zeiss)	8 m—7 km	60 MHz	5 mm $\pm$ 2$\times$10^{-6}	Mean square of series of observations
Electro-optic Pockels effect polarization modulator and demodulator	Mekometer II	0—1·5 km	500 MHz	0·3 mm $\pm$ 3$\times$10^{-6}	Single observation
	Mekometer III	0—3 km	500 MHz	0·1 mm $\pm$ 3$\times$10^{-6}	No corrections to apply
Infra-red, amplitude modulated	DI 10 Distomat Wild Heerbrugg	0—1 km	14 MHz	$\pm$ 10 — 20 mm	Mean square error Digital read-out

There are not many length measurements which can be made with the full accuracy of the standard and in many cases it is already more accurate as well as far simpler to determine length from the time of travel of electromagnetic waves. Existing equipments are listed in Table II. Future developments will probably make such methods applicable for smaller lengths and the time may arrive when it will be advantageous to define the value of c_o as a unit of measurement and make length a derived or secondary measurement.

The results given in Table I have been discussed in the text and we do not propose to derive from them a "best" value. The International Scientific Radio Union (U.R.S.I.) and the International Union for Geodesy and Geophysics (I.U.G.G.) have since 1958 recommended the international adoption of the value of the velocity of light *in vacuo* to be

$$c_0 = 299{,}792{\cdot}5 \pm 0{\cdot}4 \text{ km/sec}.$$

At the time of writing this is still the generally accepted figure.

We would repeat our philosophy of experimental measurement. The most important objective should be to increase the precision of measurement so that systematic errors can be measured and eliminated. Experience shows that extensive averaging processes invariably leave unsuspected systematic errors in the result. We see no advantage in taking a vast number of measurements as was done in the classical optical methods and in some of the recent determinations. We also regard it as unsound to take the standard deviation of the mean instead of that of a single observation as residual systematic errors are not reduced by taking more measurements. From the point of view of precision Froome's determination of 1958 is the only one to exceed those of Essen (1950) and Hansen and Bol (1950).

The frequencies of the waves used cover the very wide range of approximately 10^9 Hz to 10^{15} Hz and it can be stated that within this range the velocity is constant to 1 part in 10^6. To a much lower accuracy of about 1 % the range can be extended to 4×10^{22} Hz. Cleland and Jastram (1951) and Luckey and Weil (1952) have obtained values for the velocity of gamma rays from measurements designed to test high resolution coincidence circuits. The resolution obtained was 10^{-9} sec. Gamma rays emitted simultaneously travelled different paths to two counters and the counters were connected to the coincidence circuit in one case through a coaxial delay line. The change in the path length (about 300 m) was compensated by a coaxial delay line which was checked independently by microwave measurements of its transmission properties.

CHAPTER 12

New Methods under Consideration

1. THE N.P.L. MEKOMETER III

Although the latest version of the Mekometer (Froome and Bradsell 1967) has been designed as a high resolution distance measuring instrument for civil engineers and cadastral surveying where the interest is in ranges of less than 3 km, it is in fact potentially the most accurate modulated light-beam method yet produced for a speed of light determination. There are two reasons for this: firstly, elliptic polarization modulation is used and this has been shown to lead to a phase-measuring system which is not affected by changes of intensity and which does not suffer from the phase errors inherent in electronic systems; secondly the use of a high modulation frequency (492 MHz) has made the device extremely sensitive.

The "Mekometer III" is illustrated diagrammatically in Fig. 12.1 and its description is as follows. The light-source depicted is a high-pressure xenon flash-tube arranged to produce an intense spark of duration 1 μsec at a repetition frequency of 100 Hz. This light is plane-polarized by means of a polaroid disc. The modulator is a crystal of potassium dihydrogen phosphate (KDP), "demodulation" being achieved by a second, identical crystal. Both crystals are placed at the high impendance end of a quarter-wave tunable cavity, resonant at about 492 MHz, and pulsed into operation for 40 μsec, the optic-axis (z-axis) of each crystal being arranged to be parallel to the electric field at the end of the cavity. The light emerging from the modulator is elliptically polarized at the frequency of the driving cavity. By means of the optical system shown, this light is transmitted to a reflector (cube-corner type) located at the far end of the line to be measured. The light returned from this far-point traverses an internal variable light path (which serves to measure phase), passes through the second KDP crystal and on to the photo-multiplier. Between this KDP crystal and photo-detector is a second polaroid crossed relative to the first.

Crystals of the KDP type exhibit the linear electro-optic effect (Pockels effect) so that after passage through the second crystal the ellipticity is either enhanced or completely cancelled depending upon the instantaneous phase of the electric field relative to the phase of the light returned from the far-point. For cancellation the photo-multiplier sees a sharply defined minimum as the variable light path is moved through this state. For high modulation levels the maxima are broadened (Froome 1966) so that the

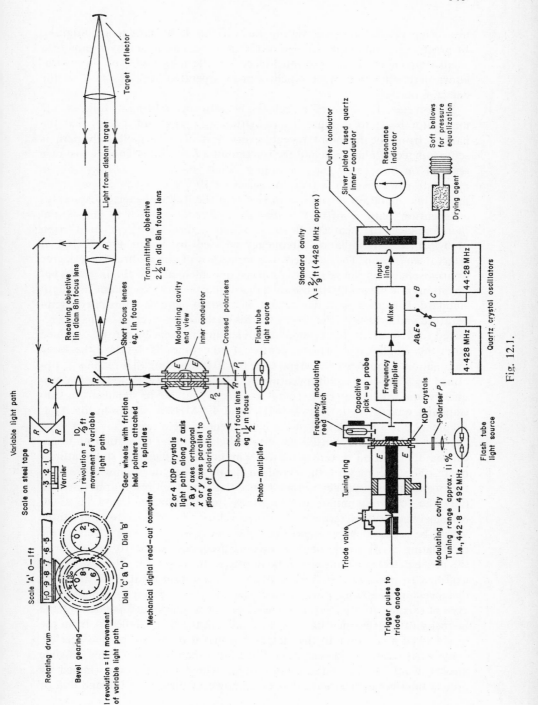

Fig. 12.1.

measuring process involves setting the variable light path to a minimum. In practice the minimum can be displayed on a null-meter by means of a small frequency modulation introduced on alternate cycles of the modulator, in conjunction with synchronously operated reed-switches in the detector output.

Mekometer III measures in feet, the actual process being as follows: the modulator is set to produce a modulation wavelength of 2 ft so that the pattern from the far-point repeats every 1 ft. The variable light-path is set to a detector minimum and the fraction of a foot in the required distance read off on the scale. The "rounding" dials shown are set to zero, the modulation wavelength is then increased by 10% and a new phase-setting made, this operation giving rise to (see also the Tellurometer, Chapter 10) a repetition pattern of 10 ft intervals. The reading on the "units" dial therefore gives the nearest 1 ft in the required distance. Next a reduction of 1% in the modulation frequency is used to obtain the 0—100 ft interval, and so on. The required distance is thus obtained directly. A degree of atmospheric refractive index compensation adequate for the intended applications is achieved by the use of a fused quartz cavity resonator as length standard. It is filled with dry air but allowed to acquire atmospheric temperature and pressure. The theory of this method was given in the papers referred to. In fact all the required modulation frequencies can be obtained from this single cavity which operates at nine times (4·4 GHz) the basic modulation frequency (492 MHz). The 0—10 ft rounding dial reading is obtained by a ten-times multiplication of the (now lower) modulation frequency, the smaller intervals being obtained by mixing-in signals from substandard quartz crystals at 44 MHz and 4·4 MHz as shown in Fig. 12.1. Figure 12.2 is a photograph of the instrument in field-use.

Even with the inexpensive flash-tube light-source a resolution of 50 μm can be obtained on distances up to 1 km, given a sufficiently stable atmosphere. The accuracy of the standard cavity is, of course, much less than the proportional accuracy needed to use this very high resolution. But with only the following minor modifications the Mekometer could make an extremely accurate determination of the velocity of light; the xenon source would be replaced by a monochromatic source such as a small laser; the modulating cavity would be driven from a frequency standard; the meter in the phase detector circuit would be improved in sensitivity. These would be expected to produce a sensitivity of phase resolution of at least 10 μm on distances up to 800 m. Furthermore, there is every reason to suppose that even at this sensitivity the phase-measurement would be independent of the intensity of the returned light, so that moving the target-reflector from the near end of a base-line to the far point would involve no "hidden" errors.

On what bases would such a modified Mekometer be used? One possible answer would be the Vaisala (1958) type whose length is determined by optical interference methods. The best known of these is in Finland and of

Fig. 12.2

length 840 m, but the Loenermark base of 576 m in Holland would also be particularly suitable. The Vaisala method for establishing these highly accurate lengths is based upon an optical multiplying system whereby the length of a 1 m etalon is multiplied (by means of fringes in white light) up to the total required distance. It is only possible to obtain one uncoloured fringe between two interfering beams of white light when the ray paths are exactly equal. Thus if, for example, an accurately known etalon of 1 m spacing between its mirrors is illuminated by a parallel beam of white light, the tenth internal reflection can be combined with a single reflection taking place within an etalon 10 m long. When a colourless dark fringe is seen the larger etalon is exactly 10 times the shorter in length. This method can be extended up to distances of several hundred metres by the use of an appropriate number of etalons and provided a suitably stable site is available. In fact the greatest difficulty is presented by atmospheric turbulence so that the fringes in white light from the largest etalon can only be observed very occasionally. In the Vaisala system the initial 1 m etalon is set-up using a spherically-ended 1 m fused quartz gauge the length of which is independently determined by optical interferometry (at one of the national standardizing laboratories) in terms of the krypton-86 wavelength standard.

For the full use of the sensitivity of the Mekometer a modification to the Vaisala system would be required: the first etalon of 1 m would have to be determined directly by monochromatic interferometry and also concurrently with the Mekometer measurement. In addition, the white-light fringe multiplication would have to be accomplished simultaneously. By these means the atmospheric effect on the Mekometer ray path would be very nearly identical with the effect on the multiplying system; the only deviation being one of refractive index difference between that appropriate for the group velocity of the Mekometer light and that of the normal refractive index of the air in the Vaisala etalons. Since the difference between group and phase velocity is rather less than 1 in 10^5 per atmosphere, temperature, pressure or humidity measurements (along the path) of quite modest accuracy would serve to correct for this difference to a residual uncertainty of less than 1 in 10^8. The effective accuracy of the total length measurement would thus approach that of the wavelength standard used in the determination of the 1 m etalon and this could be a He-Ne laser source compared at the same time with the krypton-86 standard. The measurement of the Mekometer modulation frequency could be an order of magnitude better than the length measurement, so that the expected accuracy of the overall experiment should approach 3 parts in 10^8, or 0·01 km/sec on the velocity of light.

Another, and possibly more attractive method of making the length-measurement would be by direct fringe-count of the movable reflector of a laser interferometer. Given a stable atmosphere it should be possible to count interference fringes from, say, a helium-neon laser source for distances of a few hundred metres. Mekometer III using the same type of light-source would be operated simultaneously, thus eliminating atmospheric refraction as described for the Vaisala system. Probably it would be essential to use a long, closed room for the experiment; a ship-testing tank would be particularly suitable because the displacement of the reflectors could be made by fixing them to the movable test carriage. The expected accuracy would again be $\pm 0\cdot 01$ km/sec on velocity, if the laser source used for the fringe-count was continuously compared with the Kr86 wavelength standard.

2. PROPOSED METHOD USING GAMMA RAYS

An interesting method has been suggested by Mockler and Brittin (1961) which would utilize the Mossbauer effect (1958) in a detection system suitable for the measurement of the velocity of gamma rays. It is also unusual in that it is a "single-transit" method.

Figure 12.3 illustrates the proposed system. The modulator and detector would both consist of longitudinaly oscillating quartz rods. The first (modulating) rod would have its oscillating surface covered with a thin film of

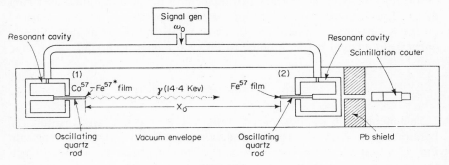

Fig. 12.3

$Co^{57}-Fe^{57*}$ which emits radiation from the well-known Mossbauer transition at an energy of 14·4 keV. The spectral quality of this transition is characterized by an exceptionally narrow band-width: for a radiation frequency of 4×10^{18} Hz the line-width is stated to be only 4 MHz, or 1 in 10^{12}.

The second quartz rod acts as detector of the frequency modulation produced by the first quartz rod, and would be covered on its oscillating surface by a thin film of Fe^{57}. The proposed separation of the two rods (X_0) was about 3 m. The frequency of the radiation emitted from the first surface will, in general, differ from the frequency "observed" by surface 2. However, when the proper time relationship exists between surfaces 1 and 2 the gamma radiation will be strongly absorbed by the second surface and this will be indicated by a drop in counting-rate of the scintillation counter.

Consider a gamma-ray emitted from surface 1 at time t_1 with frequency f_1, received at surface 2, at t_2 where the "observed" frequency is f_2, in general different from f_1. To a high degree of accuracy these frequencies are related by the equation:

$$f_1 = f_2 \left\{ 1 - \frac{a\omega_0}{c} \left[\left(\cos \frac{\omega_0 X_0}{c} - 1 \right) \sin \omega_0 t_1 + \sin \frac{\omega_0 X_0}{c} \cos \omega_0 t_1 \right] \right\}. \tag{12·1}$$

a is the amplitude of vibration of each quartz rod (assumed equal), $\omega_0 = 2\pi f_0$ where f_0 represents their frequency of vibration.

When $f_1 = f_2$ the gamma rays at surface 2 will be absorbed. Then, $\omega_0 X_0/c = 0, 2\pi \ldots 2n\pi$ where n is an integer number. If f_0 could be as high as 10 GHz, Mockler and Brittin calculate a possible amplitude as great as $a = 5 \times 10^{-8}$ cm. This would give rise to a peak velocity in excess of 100 cm/sec which is 10^5 times the minimum detectable on the basis of line width. The first detector minimum will occur when $X_0 = c/f_0 = 3$ cm, and so on.

It is suggested that because of the expected precision of determination of absorbtion the overall accuracy of this method might be as good as 1 part in 10^7. The distance X_0 could be measured by optical interferometry based on the Kr86 standard line to, say, 1 part in 10^8 and the quartz crystal frequencies (f_0) could be determined to an even higher accuracy.

3. THE LASER BEAT METHOD

This method has been proposed by Baird (1967) at the Joint Institute for Laboratory Astrophysics (Boulder, Colorado, U.S.A). It can be seen from Fig. 12.4 that the basic system is very simple in principle. A laser (which can be helium-neon or pure neon) is chosen to oscillate in two separate axial modes, the beat frequency between these modes being detected by a photodetector on to which a small proportion of the laser light falls after reflection from the semi-reflector.

For adjacent axial modes of operation of the same spectral transition the beat-frequency would be of the order of 300 MHz. Much higher frequencies can be obtained by beats between different transitions; indeed the problem for this method of operation is to find two such oscillators whose beat comes within a practical range of detection and measurement.

It is proposed to measure the difference in optical wavelength of the laser modes by means of an evacuated Fabry-Perot etalon approximately 30 m long. The accuracy of the method depends upon the very precise comparison of small wavelength differences so that a very long etalon is essential. For example, a beat frequency of 300 MHz would correspond to a fringe difference count of only 60 in this etalon. Thus for even the modest accuracy of 1 in 60,000 the precision of setting on an interference fringe would need to be 10^{-3} of an order. In fact, it is hoped to operate with two separate transitions in a pure neon laser (Hall and Morey 1967) giving a beat at 51 GHz, this would correspond to 10^4 fringes difference between the two spectral lines and the accuracy would now approach 1 in 10^7, if the centre of the fringes could be located with the same stated precision. The wavelengths of this neon doublet are near 1·1 μm.

12. NEW METHODS UNDER CONSIDERATION

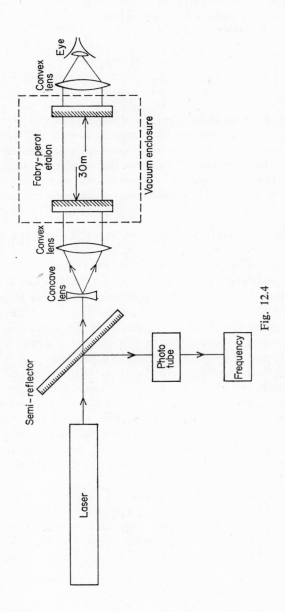

Fig. 12.4

The theory of the method is straightforward: if f_1 and f_2 represent the optical frequencies of the two laser oscillators of wavelength λ_1, λ_2, then the measured "beat" frequency is $f_1 - f_2$ where

$$f_1 - f_2 = c_0 \left(\frac{1}{\lambda_1} - \frac{1}{\lambda_2} \right)$$
$$= c_0 \left(\frac{\lambda_2 - \lambda_1}{\lambda_1 \lambda_2} \right). \tag{12.2}$$

$f_1 - f_2$ can be measured to an accuracy of better than 1 in 10^8; λ_1 and λ_2 can each be measured to 1 in 10^8 by comparison with the Kr86 wavelength standard. There is no refractive index correction to consider because it is convenient to use an evacuated etalon for the wavelength difference $(\lambda_2 - \lambda_1)$ comparison. Thus the accuracy of the velocity determination is basically limited by the precision of this wavelength difference measurement.

4. THE LIGHT-PULSE RECYCLING OSCILLATOR

This method has been proposed by Gerharz (1957) and is based upon the possibility of generating flashes of light of extremely short duration from a specially treated dynode of a photomultiplier tube. This pulse of light can be directed upon a distant mirror and returned to the cathode of the phototube where it gives rise to a burst of electrons. The electrons will be multiplied greatly upon traversing the tube until they arrive at the fluorescent dynode when another pulse of light will be generated.

Figure 12.5 illustrates diagrammatically a possible form of this recycling oscillator suitable for the measurement of the velocity of light. The final dynode of the photo-multiplier tube, P, is coated with magnesium oxide which can give off a pulse of blue light of duration 10nsec (10^{-8} sec) when an electron cloud hits it. The emitted light would be collimated by the lens, L, and directed upon the movable mirror, M. M could also be part of an optical interferometer so that the displacement of it could be assessed highly accurately by fringe counting.

It is seen that the velocity of light can be obtained from a measurement of the change of recycling frequency as the mirror M is displaced through a distance of, say, one metre. If F is the original repetition frequency and f the new repetition frequency after the optical path has been increased by δs, Gerharz shows that the group velocity of light in air is given by:

$$c = \frac{2f\delta s}{1 - f/F}.$$

One of the principal difficulties of operating this method in practice is that of obtaining sufficiently uniform electron transit-time through the photo-multiplier, but Gerharz considers that given a very stable high tension supply to the tube the statistical fluctuations due to differing electron trajectories should average out. The light-pulse emitted by the dynode is not of great intensity so that the overall optical path would appear to be limited to a few metres. Also, the change in intensity of the pulse returned to the photo-tube cathode will produce an error if the optical displacement becomes too large.

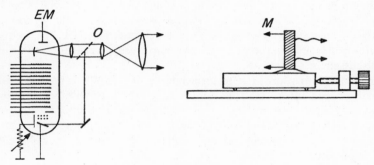

Fig. 12.5.

Gerharz considers that if these difficulties could be overcome the method might lead to a value for the velocity of light accurate to a few parts in 10^7 for a mirror displacement in the region of 1 m.

5. SUB-MILLIMETRE WAVE INTERFEROMETER

In the years following the 72 GHz microwave interferometer described in Chapter 8, considerable progress has been made towards extending the range of microwave frequency measurements into the sub-millimetre region of wavelengths. Froome (1964) has generated harmonics from a 35 GHz Klystron oscillator and detected the 29th corresponding to a wavelength of 0·29 mm with a frequency of 10^{12} Hz. Each harmonic is potentially as good in quality as the fundamental. Within the next few years it would seem quite within the bounds of possibility to measure the frequencies of infra-red lasers in the 0·1 — 0·01 mm wavelength region, for it is well established that millimetre-wave electronic oscillators can be stabilized in frequency to better than 1 part in 10^8; thus harmonic generation, or harmonic mixing from such a source with an infra-red laser, is a very powerful method of frequency measurement.

To obtain a velocity determination the wavelength of the sub-millimetre source would have to be measured extremely accurately. A microwave interferometer operating in the Fresnel diffraction region would be the best method for accomplishing this. An interferometer 50 m in length would permit a maximum order of interference of 10^6 half-waves from a 0.1 mm source. The diameter of the reflectors could be less than 30 cm, so the whole interferometer could be evacuated. In the Fresnel region the residual uncertainty of assessing the diffraction correction (expressed as a fraction of a wavelength) is greater than for the Fraunhofer region described in Chapter 8. However, analysis shows that it should be possible to obtain the diffraction correction to better than 1% of a fringe spacing or about 1 in 10^8 on the wavelength measurement. The spacing of the interferometer plates could be obtained by use of an optical laser whose wavelength would be compared with the Kr86 standard — also to an accuracy of 1 in 10^8. The frequency measurement would be rather better. An improved cavity resonator method was discussed in Chapter 6.

Thus a determination of c_0 to an accuracy of 2 in 10^8 is probable in the reasonably near future. If this comes to fruition it will be tempting to suggest defining a value of the speed of light by international agreement and then deriving the standard of length from such an interferometer *via* the use of frequency measurement which is the most accurately realized of all standards.

REFERENCES

Anderson, W. C. (1937) *Rev. sci. Instr.*, 8, 239.
Anderson, W. C. (1941) *J. Opt. Soc. Am.*, *31*, 187.
Aslakson, C. I. (1949a), *Nature (Lond.)*, *64*, 711.
Aslakson, C. I. (1949b), *Trans. Am. geophys. Union*, 30, 475.
Aslakson, C. I. (1951) *Proc. Am. Soc. Civil Eng.*, 77 Separate No. 52.
Baird, R. C. (1967) *Proc. IEEE*, 55, 1032.
Barrell, H. and Puttock, M. J. (1950) *Br. J. Appl. Phys. 1*, 87.
Barrell, H. and Sears, J. E. (1939) *Phil. Trans. R. Soc. (Lond.)* A, *238*, 1.
Bearden, J. A. and Watts, H. M. (1951) *Phys. Rev.*, *81*, 73.
Bedard, F. D. Gallagher, J. J. and Johnson, C. M. (1953) *Phys. Rev.*, *92*, 1440.
Bergstrand, E. (1949a) *Nature (Lond.)*, *163*, 338.
Bergstrand, E. (1949b) *Ark. Mat. Astr. Fys.* 36A, 1.
Bergstrand, E. (1950) *Ark. Fys.* 2, 119.
Bergstrand, E. (1951) *Ark. Fys.* 3, 479.
Bergstrand, E. (1957) *Ann. Franc. Chronom.*, *11*, 97.
Bernier, J. (1946) *Onde Elect.* 26, 305
Birge, R. T. (1941) *Rep. Phys. Soc. Progr. Phys.*, 8, 90.
Bleaney, B. I. and Bleaney, B. (1957) "Electricity and Magnetism", Oxford, Clarendon Press.
Blondlot, R. (1891) *C. r. hebd. Acad. Sci. Paris*, *113*, 628
Bol, K. (1950) *Phys. Rev.* 80, 298.
Bradley, J. (1728) *Phil. Trans. R. Soc. (Lond.)*, 35, 637, p.308, in collected volumes 1724—1734.
Cleland, M. R. and Jastram, P. S. (1951) *Phys. Rev.* 84, 271.
Cornu, A. (1874) *J. l'Ecole polytechn.* 27, (44), 133.
Debye P. (1929) *Polar Molecules* (New York: Chemical Catalogue Co)
Dorsey, N. E. (1944) *Trans. Am. phys. Soc.*, 33, 1.
DuMond, J. W. M. and Cohen, E. R. (1948) *Rev. mod. Phys*, 20, 82.

Edge, R. C. A. (1956) *Nature (Lond.)*, 177, 618.
Edlén, B. (1953) T. *Opt. Soc. Am.*, 43, 455.
Essen, L. and Froome, K. D. (1951) *Proc. Phys. Soc, B*, 64, 862.
Essen, L. (1946) *J. IEE* 93, Part III A, 1413.
Essen, L. (1947) *Nature (Lond.)* 159, 611.
Essen, L. (1950) *Proc. R. Soc. (Lond.)* A 204, 260.
Essen, L. and Gordon-Smith, A. C. (1945) *J. IEE*, 92, Part III A, No. 9, 1374.
Essen, L. and Gordon-Smith, A. C. (1948) *Proc. R. Soc. (Lond.)* 194 A, 348.
Essen, L. and Parry, J. V. L. (1957) *Phil. Trans. R. Soc. Lond. A.*, 250, 45.
Fizeau, H. (1849) *C. r. hebd. Acad. Sci. Paris*, 29, 90, 132.
Florman, E. F. (1955) *J. Res. Natl. Bur. Stand.*, 54, 335.
Foucault, L. (1862) *C. r. hebd. Acad. Sci. Paris*, 55, 501, 792.
Froome, K. D. (1952) *Proc. R. Soc. (Lond.)*, 213 A, 123.
Froome, K. D. (1952) *Nature (Lond.)*, 169, 107.
Froome, K. D. (1954) *Proc. R. Soc. (Lond.)*, 223 A, 195.
Froome, K. D. (1955) *Proc. Phys. Soc. (Lond.)*, 68 B, 883.
Froome, K. D. (1958) *Proc. R. Soc. (Lond.)*, 247 A, 109.
Froome, K. D. (1964) *Quantum Electronics* III, 2. Columbia University Press.
Froome, K. D. (1966) *J. Sci. Inst.*, 43, 129.
Froome, K. D. and Bradsell, R. H. (1967) *Proceedings of Commonwealth Survey Officer's Conference, Cambridge* 1967. H. M. Stationery Office, London.
Gerharz, R. (1957) *J. Electron.*, 2, 416.
Gilliam, O. R., Johnson, C. M., and Gordy, W. (1950) *Phys. Rev.*, 78, 140.
Hall, J. L. and Morey, W. W. (1967) *Appl. Phys. Lett.*, 10, 152.
Hartshorn, L. (1947) "Radio Frequency Measurements", London, Chapman and Hall.
Hertz, H. (1887) *Wiedemann Ann. Phys. Chem,.* 31, 421.
Houstoun, R. A. (1950) *Proc. R. Soc., Edinburgh*, 63, Part 1, 95.
Hüttel, A. (1940) *Ann. Phys. Ser.* 5, 37, 365.
Jones, F. E. (1947) *JIEE*, 94, Part 3, 399.
Jones, F. E. and Cornford, E. C. (1949) *J. IEE*, 96, Part 3, 477.
Karolus, A., *Fifth International Conference on Geodetic Measurement*, 1965, Deutsche Geodetische Kommission, Munich, 1966, p. 1.
Karolus, A. and Mittelstaedt, O. (1928) *Phys. Z.*, 29, 698.
Karolus, A. and Helmberger, J. (1964) *Ann. Phys.*, Ser. 7, 14, 36.
Luckey, D. and Weil, J. W. (1952) *Phys. Rev.* 85, 1060.
Mackenzie, I. C. C. (1954) "The Geodimeter measurement of the Ridge Way and Caithness Bases". *Ordnance Survey Professional Papers, New Series*, No. 19, H. M. Stationery Office, London, 1954.
Maxwell, J. C., (1904) "A Treatise on Electricity and Magnetism", Vol. II, Oxford, Clarendon Press.
McKinley, D. W. R. (1950) *R. Astron. Soc. ,Canada*, 44, 89.
Mercier, J. (1923) *Ann. Phys. Ser.* 9, 19, 248, 20, 5.
Mercier, J., (1924) *J. Phys. Radium*, (6), 5, 168.
Michelson, A. A. (1880) *Astron. Papers Am. Ephemeris*, 1, 109.
Michelson, A. A. (1891) *Astron. Papers Am. Ephemeris*, 2, 231.
Michelson, A. A. (1927) *Astrophys. J.*, 65, 1.
Michelson, A. A., Pease, F. G. and Pearson, F. (1935) *Astrophys. J.*, 82, 26.
Mittelstaedt, O. (1929) *Ann. Phys. Ser.* 5, 2, 285.
Mockler, R. C. and Brittin, W. E. (1961) *Natl. Bureau Stand. (Boulder, Colorado)* Rpt No. 6762.
Mossbauer, R. L. (1958) *Z. Phys.* 151, 124.
Mulligan, J. F. (1952) *Am. J. Phys.*, 20, 165.
Nethercot, A. H., Klein, J. A. and Townes, C. H. (1952) *Phys. Rev.*, 86, 798.
Newcomb, S. (1891) *Astron. Papers Am. Ephemeris*, 2, 107.

Newton, Sir Isaac, (1931) "Optiks", London, Bell and Sons.
Plyler, E. K., Blaine, L. R. and Connor, W. S. (1955) *T. Opt. Soc. Am. 45*, 102.
Rank, D. H., Bennett J. M. and Bennett H. E. (1956) *J. opt. Soc. Am., 46*, 477.
Rank, D. H., Ruth, R. P. and Vander Sluis, K. L. (1952) *J. opt. Soc. Am., 42*, 693.
Rank, D. H., Shearer, J. N., and Wiggins, T. A. (1954) *Phys. Rev., 94*, 575.
Rosa, E. B. and Dorsey, N. E. (1907) *Bull. US, Bureau of Standards, 3*, 433.
Ross, J. E. R. (1951) *Union of Geodesy and Geophysics, 9th General Conference.*
Ross, J. E. R. (1954) *Union of Geodesy and Geophysics, 10th General Conference.*
Schelkunoff, S. A. (1939) *Phys. Rev. 56*, 308.
Schöldström, R. (1955) "Determination of Light Velocity on the Oland Base Line 1955" (Issued by AGA Ltd., Stockholm).
Simkin, G. S., Lukin, I. V., Sikora, S. V. and Strenlenskii, V. E. (1967) *Izment, Tech. 8*, 92.
Slater, J. C. (1950) "Microwave Electronics", New York, Von Nostrand Co. Inc.
Smith, R. A., Franklin, E. and Whiting, F. B. (1947) *JIEE, 94*, Part III, 391.
Smith, E. K., and Weintraub, S. (1953). *Proc. I. R. E.*, 41, 1035
Vaisala, Y. (1958) *Handbuch Vermessungskunde (Stuttgart) 4*, 482.
Van Vleck, I. H. (1942) MIT *Rad. Lab. Rep. No* 43.2
Wadley, T. L. (1957) *Empire Survey Review, 14*, Nos. 105—106, 100.
Wadley, T. L. (1958a) *Empire Survey Rewiev, 14*, Nos. 107—109, 227.
Wadley, T. L. (1958b) *Trans. South Africa IEE*, May 1958.
Zacharias, J. R. and Harrison, G. R. (1956) RLE(MIT) *Q. Prog. Rep.*, Jan.

GENERAL BIBLIOGRAPHY

Bearden, J. A. & Thomson, J. S. (1955) "A Survey of Atomic Constants". John Hopkins University, Baltimore
Cohen, E. R. and Du Mond, J. W. M. (1955) *N. /m. Sc. Centre Monograph, 182*
Du Mond, J. W. M. (1959) *Annals of Physics* 1 365
Du Mond, J. W. M. and Cohen, E. R. (1953) *Rev. Mod. Phys.* 25 691
Essen, L. (1952) *Science Progress* 157, 54
Essen, L. (1956) *Endeavour* 15 No. 58, 87
Froome, K. D. (1956) *J. Brit. I. R. E.*, 16, 497
Sanders, J. H. (1965) "The Fundamental Atomic Constants". Oxford Univ. Press.
Sanders, J. H. (1965) "The Velocity of Light". Pergamon Press, Oxford

Author Index

A

Airy, G. B. 22
Ampère A. M. 7
Anderson, W. C. 31, 36, 49
Aslakson, C. I. 75, 76, 78, 79, 137

B

Bauman, Col. 128
Babinet J. 21
Barclay 72
Baird, R. C. 146
Barrell, H. 25, 26, 67
Bedard, F. D. 83
Bearden, J. A. 72, 73
Benoit, A. 21
Bennett, J. M. 82, 137
Bergstrand, E. 31, 75 114, 120, 122, 137
Bernier, J. 55
Birge, R. T. 29, 114
Blaine, L. R. 83, 137
Bleaney, B. 46
Bleaney, B. I. 46
Blondlot, R. 8, 10, 41
Bol, K. 72, 73, 137, 139
Bradley, J. 2, 9
Bradsell, R. H. 140, 142
Brittin, W. E. 145
Bennett, H. E. 18, 137

C

Cauchy, A. L. 27, 28
Cleland, M. R. 139
Cohen, E. R. 152
Connor, W. S. 83, 137
Cornford, E. C. 79, 137
Cornu, A. 5, 9
Coulomb, C. A. de, 6

D

Debye, P. 25
Delambre J. B. J. 1

Dorsey, N. E. 6, 29, 41, 49
Duane, L. W. 10
Du Mond, T. W. M. 152

E

Edge, R. C. A. 122, 132
Edlén, B. 27
Essen, L. 23, 26, 50, 55, 57, 63, 64, 73, 74, 75, 79, 89, 95, 114, 133, 136, 137, 139
Evelyn 20

F

Fabry, C. 10
Faraday, M. 7, 8, 13, 37
Fizeau, H. 3, 9, 30, 34
Florman, E. F. 111, 113, 137
Foucault, L. 4, 9, 30
Franklin, E. 78
Fraunhofer J. von, 90
Fresnel 12, 18, 90, 149
Froome, K. D. 26, 79, 87, 93, 95, 97, 100, 110, 133, 136, 137, 140, 149

G

Galileo, G. 1, 22
Gallagher, J. J. 83
Gerharz, R. 148
Giebe, E. 23
Gilliam, O. R. 83
Glasenapp 1
Glazebrook, R. 14
Gordon-Smith, R. C. 50, 55, 57, 114, 137
Gordy, W. 83
Graham 20, 22
Grimaldi, C. C. 18

H

Hall, J. L. 146
Halley, E. 22

AUTHOR INDEX

Hansen, W. 72, 73, 137, 139
Harrison, G. R. 74
Hartshorn, L. 47
Helmberger, J. 125, 126
Hertz, H. 8, 17
Hooke, R. 12, 18
Houstoun, R. A. 84, 137
Hüttel, A. 31, 40, 49
Huygens, C. 12, 104

J

Jastram, P. S. 139
Johnson, C. M. 83
Jones, F. E. 79, 137

K

Karolus, A. 30, 33, 114, 125, 136, 137
Klein, J. A. 81
Kohlrausch, L. 7, 10, 41

L

Lorentz, H. A. 13
Lorenz, L. 43
Luckey, D. 139
Lukin, I. V. 113, 137, 152

M

Mackenzie, I. C. C. 122
Maxwell, J. C. 7, 10, 13, 16, 42, 45, 46
McKinley, D. W. R. 85, 137
Mercier, J. 45, 47, 49, 50, 51
Michelson, A. A. 6, 9, 21, 30, 31, 32, 49
Mie, G. 45
Mittelstaedt, O. 30, 33, 49
Mockler, R. C. 145
Mossbauer, R. L. 145
Morey, W. W. 146

N

Nethercot, A. H. 81
Newcomb, S. 6, 9, 30
Newton, I. 12

P

Parry, J. V. L. 23
Pearson, F. 30, 32, 49
Pease, F. G. 30, 32, 49
Perot, J. B. G. G. 10
Plyler, E. K. 83, 137

Pound, R. V. 97
Puttock, M. J. 67

R

Rank, D. H. 81, 95, 137
Rayleigh, Lord 50
Rive, La. 9
Roemer 1, 11, 22
Rosa, E. B. 10, 41, 49
Ross, J. E. R. 79
Ruth, R. P. 81, 137

S

Sanders, J. H. 152
Sarasin, E. 9
Scheibe, A. 23
Schelkunoff, S. A. 104
Schöldström, R. 137
Sears, J. E. 25, 26, 27
Shearer, J. N. 81, 137
Sikora, S. V. 113, 137, 152
Simkin, G. S. 113, 137, 152
Shortt, W. H. 22
Slater, J. C. 57
Smith, R. A. 78, 137
Sommerfeld, A. 45
Strenlensku, V. E. 113. 137, 152

T

Thomson, J. J. 10, 50
Tompion, T. 22
Townes, C. H. 81

V

Vander Sluis, K. L. 81, 137
Van Vleck, I. H. 25

W

Wadley, T. L. 128, 129, 132, 133, 137
Watts, H. M. 72, 73
Weber, W. E. 7, 10, 41
Weil, U. W. 139
Whiting, F. B. 78
Wiggins, T. A. 81, 137

Y

Young, T. 12

Z

Zacharias, J. R. 74

Subject Index

A

Aberration of light 2, 9
Accuracy of measurement 20, 23, 28, 29, 57, 79, 81, 110, 136
Assman hygrometer 89
Atmospheric turbulence 144
Atomic clock 23
Atomic energy levels 80
Attenuator, constant phase 100, 101
Attenuator, variable 87

B

Base line, Caithness 122
Base line, Linkoping 120
Base line, Loenermark 143
Base line, Oland 121
Base line, Ridge Way 122
Base line, Vaisala 127
Beam divider 97, 127
Bessel functions 51, 52

C

Carnegie Corporation 32
Cauchy equation 27, 28
Cavity resonator, frequency of 51
Cavity resonator method 50
Cavity resonator, modes of resonance 54
Cavity resonator, quality factor of 55, 56
Cavity resonator refractometer 95
Cavity resonator, theory of 51
Circuit, tuned, electrical 8, 25
Clocks, atomic 23
Clocks, pendulum 22, 44, 45
Clocks, quartz 22
Clocks, tuning fork 22
Conductivity 16, 46, 55, 57
Constant phase waveguide interferometer (c. p. i.) 100, 101
Corpuscular theory of light 12
Corner reflectors, cube 124

D

Debye equation 25
Delay line, electrical 118, 120
Delay line, optical 118, 120, 140
Detector, crystal 59
Detector, photoelectric cell 36, 38, 39
Detector, superheterodyne 59, 88
Diffraction error 87, 88, 90, 95, 102
Diffraction grating 81, 84, 126
Diffraction in Fraunhofer region 90
Diffraction of light 12, 18
Dispersion of light 13
Doppler effect 1
Distance measurement 75, 114, 128, 140
Draconis 2

E

Electric field 15
Electrical methods 6, 10, 41
Electrical units 7, 15, 41
Electro-optic effect of Kerr cell 33, 115
Electromagnetic theory of light 8, 15, 50
Electromagnetic radiation 8, 9
Electromagnetic waves 17
End standards of length 88, 96, 99
Error; setting 23, 57
Error; systematic 24, 28, 41, 57, 139

F

Fabry-Perot interferometer 86
Field equations 16
Force between charges 6, 15
Force between magnetic poles 7, 15
Fraunhofer diffraction region 90, 95, 125
Frequency of cavity resonators 51
Frequency modulation 129
Frequency of repetition 4
Frequency locking 45, 87
Frequency measurement 4, 32, 33, 35, 44, 45, 50, 59, 62, 67, 86, 89, 97
Fresnel integrals 18

G

Gamma rays 18, 139, 145
Gauges, end 88, 96, 99
Gauges, slip 60, 64, 88, 99
Geodetic base line 23, 121, 122, 123, 127 143
Geodimeter 114, 138
Geodetic survey 32, 75, 76, 128
Group refractive index 27
Group velocity 14, 27, 36, 144

H

Heterodyne detector 59
Heterodyne method of frequency measurement 45, 50, 59
Horns, transmitting and receiving 87, 89, 91, 96
Huygens principle 104
Hybrid T junction 87, 97

I

Inductance of parallel wires 45, 47
Infra-red measurements 81, 83
Interferometer, Fabry-Perot 86
Interferometer, Michelson 86
Interferometer, four horn 95
Interferometer, microwave 86, 95
Interferometer, optical 21, 60, 67, 72, 86, 94, 127, 136, 142
Interferometer, radio wave 86, 111
Interferometer, Vaisala 127, 142, 143, 144
International Bureau of Weights and Measures 21

J

Jupiters satellites 1, 9

K

KDP crystal 142
Kerr cell 30, 33, 35, 36, 39, 114, 115
Klystron Pound stabilised 87, 97
Klystron Q-band 97
Krypton light source 21, 23

L

Laser 22, 86, 144, 146
Lecher wires 45

Length, standard of 20, 37, 94
Light chopper 4, 6, 30, 32, 39
Light, corpuscular theory of 12
Light, diffraction of 18, 87, 90, 102
Light, dispersion of 13
Light, electromagnetic theory of 15
Light modulator, ultrasonic 83, 125
Light, modulated beam of 30, 33, 36, 39, 83, 114, 125, 140
Light, nature of 12, 18
Light path, variable 118, 120
Light, polarised beam of 34, 118, 140
Light pulse oscillator 149
Light quanta 13
Light shutter 1, 39
Light source, mercury 39, 94, 124
Light source, spark discharge 8, 17
Light source, isotopic 21, 23
Light source, xenon flash tube 140
Light, wave theory of 12

M

Maxwell bridge 42
Magnetic field 15
Mekometer 125, 138, 142, 144
Microwave interferometer 86, 91, 96, 149
Microwave measurements 50, 64, 72, 74, 86, 95, 128
Mirror, rotating 4, 6, 30, 32
Modulator 30, 33, 36, 39, 83, 114, 125, 140
Mossbauer effect 145
Mount Wilson observatory 32

N

National Bureau of Standards (NBS) 26
National Physical Laboratory (NPL) 26, 27, 50, 89
Nicol prism 34, 39
Nitrobenzene 33

O

Oboe radar 79
Optical interferometry 60, 67, 72, 86, 94, 127, 136, 142, 143, 144

P

Pendulum clocks 21, 22, 32, 33, 44, 45
Permeability 15, 17
Permittivity 15, 17

SUBJECT INDEX

Phase measurement 129, 140
Phase shifter 97
Phase velocity 14
Photoelectric cell 36, 39, 85
Photographic zenith tube 22
Photomultiplier 39, 115, 140
Pockels' effect 140
Polarised light 34, 118, 140
Potassium dihydrogen phosphate (KDP) 140
Precision of setting 24

Q

Quality factor 55, 56, 57, 67, 70
Quartz as diffraction grating 84
Quartz cavity resonator 74, 142
Quartz clocks 23
Quartz modulator 75, 83
Quartz oscillator 22

R

Radar method 75
Radio wave interferometer 86
Ratio of em and es units 6, 7, 10, 41
Reflectors, aluminium 88
Reflections, stray 88
Refractive index of air 24, 30, 36, 89, 95, 127, 133
Refractive index of water vapour 25, 113
Refractometer, cavity resonator 95, 97
Resonant circuit 8, 45, 50
Results, summary of 10, 11, 49, 136
Rockfeller Foundation 32
Rotating mirror method 4, 5, 9

S

Scattering of radiation 89
Setting accuracy 23, 24
Shoran radar 74, 76
Silicon crystal distorter 97
Silicon crystal harmonic generator 97
Skin depth 46, 47, 55, 65, 71, 72
Spark discharge source 8, 17
Spectral lines of CO 82
Spectral lines of HCN 81
Spectroscopic method 76, 80
Standard, definitive 20
Standard deviation 28
Star transits 22
Standard of length 20
Standard of time 22

Stroboscope 5, 32
Superheterodyne detector 59, 86, 88
Surveying, optical 114, 138
Stationary waves 9, 17
Systematic errors 24

T

Tapes, invar measuring 23, 37, 48, 111, 114, 121, 133
Telescope 2
Telescope, transit 22
Tellurometer 125, 128, 138
Time standard 22, 36
Time signals 23, 33, 44
Toothed wheel method 4, 5, 9
Triangulation 76, 114
Trialateration 114
Tuning Fork 6, 22, 32, 35

U

Ultrasonic light modulator 83, 125
Ultrasonic waves 83, 114
Unit charge 6, 15
Unit magnetic pole 7, 15
Unit of resistance 43
Units, cgs electrostatic 6
Units, cgs electromagnetic 7
Units, MKS system 15
Units, ratio of 7, 10, 11, 41
University of Chicago 32
U. S. Naval Observatory 22
U. S. Coast and Geodetic Survey 32

V

Velocity of light, summaries of values 10, 11, 49, 137
Velocity of light, accepted value 146

W

Water vapour absorption 25
Wave electromagnetic 17, 18
Wave equation 13, 17, 51
Wave impedance 17
Wave polarised 34, 118, 140
Wave front 23, 102, 111
Wave, plane 24
Wave, progressive 13
Wave, stationary 9
Wave theory of light 12, 13
Wheel, toothed 4, 5, 9
Wires, parallel 9

Date Due

TN: 204855 Pieces: 1 IL: 4800283 FFL 05/02/04		
~~MAR 3 1982~~		
MAY 12 1988		
~~FEB 19 1992~~		